ECONOMIC COMMISSION FOR EUROPE

FINANCING ENERGY EFFICIENCY
INVESTMENTS FOR CLIMATE
CHANGE MITIGATION PROJECT

INVESTOR INTEREST
AND CAPACITY
BUILDING NEEDS

UNITED NATIONS
New York and Geneva, 2009

NOTE

The designations employed and the presentation of the material in this publication do not imply the expression of any opinion whatsoever on the part of the Secretariat of the United Nations concerning the legal status of any country, territory, city or area, or of its authorities, or concerning the delimitation of its frontiers or boundaries. In particular, the boundaries shown on the maps do not imply official endorsement or acceptance by the United Nations.

Mention of any firm, licensed process or commercial products does not imply endorsement by the United Nations.

ECE/ENERGY/72

UNITED NATIONS PUBLICATION
Sales No. E.09.II.E.17
ISBN 978-92-1-117013-9
ISSN 1014-7225

ii

Increasing energy efficiency globally is one of the most promising ways to tackle climate change. The countries of South-Eastern Europe, Eastern Europe and Central Asia are challenged with numerous economic and environmental problems caused by their inefficient and polluting energy systems. However, these problems present an opportunity for a significant increase in energy efficiency and reduction of greenhouse gas emissions.

Twelve countries of the region – Albania, Belarus, Bosnia and Herzegovina, Bulgaria, Croatia, Kazakhstan, Republic of Moldova, Romania, Russian Federation, Serbia, the former Yugoslav Republic of Macedonia, and Ukraine – participate in the Financing Energy Efficiency Investments for Climate Change Mitigation Project, which UNECE began implementing in January 2008. The Project aims to assist participating countries to enhance their energy efficiency and reduce air pollution and greenhouse gas emissions in order to meet international obligations under the United Nations Framework Convention on Climate Change (UNFCCC) and UNECE environmental conventions. The Project intends to facilitate appropriate policy reforms in the participating countries, create a network of energy managers to exchange experience on best practices, establish a public-private partnership investment fund with a target capital of € 250 million and develop a pipeline of energy efficiency and renewable energy projects to be financed by it.

This publication is the result of assessment missions to the participating countries, which were one of the first major steps in implementation of the Project. It provides initial assessment of public and private sector investor interest in the future investment fund and of the local financial environment. It contains an appraisal of capacities of local experts for development of investment projects in energy efficiency and renewable energy, their capacity building needs and preliminary review of national energy policy information. The report will serve as the basis for the regional analysis of policy reforms to promote energy efficiency and renewable energy investments and as a tool for decision-makers to promote energy efficiency and renewable energy policies in their countries.

Ján Kubiš
Executive Secretary and Under Secretary-General

TABLE OF CONTENTS

ACRONYMS AND ABBREVIATIONS

BEERECL	Bulgarian Energy Efficiency and Renewable Energy Credit Line
BgEEF	Bulgaria Energy Efficiency Fund
BREECL	Bulgarian residential Energy Efficiency Credit Line
CDM	Clean Development Mechanism
CHP	Combined heat and power
DNA	National designated authority
EBRD	European Bank for Reconstruction and Development
EE	Energy efficiency
EIB	European Investment Bank
EP	Electroprivreda
EPBIH	Electricity Company of Bosnia and Herzegovina
EPEEF	Environmental Protection and Energy Efficiency Fund
EPHZHB	Electricity Company of Herzeg-Bosnia
EPRS	Electricity Company of Republika Srpska
ERA	Electricity Regulatory Authority
ESCO	Energy service company
EU	European Union
FDI	Foreign direct investment
FREE	GEF Energy Efficiency Financing Facility Project
GEF	Global Environment Facility
GOF	Global Opportunities Fund
HBOR	Croatian Bank for Reconstruction and Development
IEA	International Energy Agency
IFC	International Financial Corporation
IFI	International financial institution
IPO	Initial public offering
IPA	Instrument for Pre-Accession Assistance
JI	Joint Implementation
LEF	Local Enterprise Facility
MEED	Ministry of Economy, Energy and Development
MOFTER	Ministry of Foreign Trade and Economic Relations
NC	National coordinators
NPI	National participating institutions
NPP	Nuclear power plant
NGO	Non-governmental organization
OECD	Organization for Economic Co-operation and Development
PPP	Purchasing power parity
RAEF	Romanian American Enterprise Fund
RES	Renewable energy sources
RIEEC	Romanian Industrial Energy Efficiency Company
SEEA	Serbian Energy Efficiency Agency
SHP	Small hydropower
SME	Small and medium-sized enterprises
TPES	Total primary energy supply
TSO	Transmission system operator
UNDP	United Nations Development Programme
UNFCCC	United Nations Framework Convention for Climate Change
USAID	United States Agency for International Development
SEAF	Small Enterprise Assistance Funds
UKEEP	Ukraine Energy Efficiency Programme
VAT	Value added tax

INTRODUCTION

This Investor Interest and Capacity Building Needs Report has been prepared in the framework of the Financing Energy Efficiency and Renewable Energy Investments for Climate Change Mitigation project.

The Financing Energy Efficiency and Renewable Energy Investments for Climate Change Mitigation project assists the participating countries[1] to enhance their energy efficiency, diminish fuel poverty and reduce air pollution such as greenhouse gas emissions in order to meet international environmental treaty obligations under the United Nations Framework Convention on Climate Change (UNFCCC) and the United Nations Economic Commission for Europe (UNECE).

It aims to provide a pipeline of new and existing projects dedicated to public-private partnership investment funds. It will establish an expanded and enhanced network of selected municipalities and energy managers linked by advanced Internet communications with international partners for value added information transfers on policy reforms, financing and energy management. It will provide case study investment projects in renewable energy technologies, electric power and clean coal technologies.

The project is intended to (a) identify and develop investment projects and strengthen capacities of a network for the development of energy efficiency projects; (b) provide assistance to municipal authorities and national administrations to introduce economic, institutional and regulatory reforms needed to support these investments projects; and (c) support banks and commercial companies to invest in these projects through professionally managed investment funds. It promotes a self-sustaining investment environment for cost-effective energy efficiency projects for carbon emissions trading under the UNFCCC Kyoto Protocol.

Mandate

The Ad Hoc Group of Experts on Energy Efficiency Investments for Climate Change Mitigation is the governing body of the project. At its meeting in February 2008 the Group of Experts agreed that expert missions had to be organized to each participating country for consultations on the investment fund with relevant government authorities and financing institutions to assess the local financing environment, appraise investment project development capacities and analyse the conditions under which an investment fund could operate in their country.

Objectives of the assessment missions

The assessment missions to the participating countries had the following objectives:

- Initial assessment of public and private sector investor interest in the Eastern European Energy Efficiency Investment Fund
- Preliminary analysis of the local financial environment and conditions under which an equity and mezzanine fund could operate in each country, including equity participation in energy service companies, special purpose project companies or similar entities

---

[1] The participating countries of the project include Albania, Belarus, Bosnia and Herzegovina, Bulgaria, Croatia, Kazakhstan, Republic of Moldova, Romania, Russian Federation, Serbia, the former Yugoslav Republic of Macedonia, and Ukraine.

- Appraisal of the energy efficiency and renewable energy investment project development capacities of local experts and capacity building needs

- Consultation with the National Coordinators (NCs) and National Participating Institutions (NPIs) on the assistance to be provided for the investment fund design, regional analysis of policy reforms to promote energy efficiency and renewable energy investments, and website and internet communications

- Preliminary review of national energy policy information relevant for the Regional Analysis

- Presentation of the methodology and data requirements of national case studies to be undertaken by the NPI in relation to the Regional Analysis

- Presentation of a draft UNECE–NPI Memorandum of Understanding prepared in accordance with the Revised Work Plan for the First Year of Project Operations (2008).

Participants

The Investor Interest and Capacity Building Needs Report is based on the results of assessment missions to participating countries, as well as information provided by the NCs and NPIs and other available information. It is an outcome of the joint work of the team of national and international experts.

The Project Management Unit within UNECE secretariat, mandated by the Ad Hoc Group of Experts on Energy Efficiency Investments for Climate Change Mitigation, was in charge of organizing the assessment missions. Representatives of the secretariat, including the Regional Adviser on Energy, participated in the assessment missions coordinating the arrangements required for successful implementation of the assignment. The assessment missions team included a Senior Energy Policy Adviser representing the Agency for Environment and Control of Energy (Agence de l'environnement et de la maîtrise de l'énergie – ADEME, France) and a Monitoring and Evaluation Adviser for the United Nations Foundation, one of the project donors (as part of the overall monitoring of the project operations). The Project Management Unit engaged the services of four consulting companies to conduct the assessment of investor interest and appraise capacity building needs. Representatives of these consulting companies (listed below along with participating countries they covered) were part of the assessment missions team.

Company	Participating Countries
Econoler International (Canada)	Bosnia and Herzegovina, Croatia, Serbia
Energy Saving International AS (Norway)	Albania, Kazakhstan, the former Yugoslav Republic of Macedonia,
International Consulting on Energy (France)	Bulgaria, Republic of Moldova, Romania
Renaissance Finance International Ltd. (United Kingdom)	Belarus, Russian Federation, Ukraine

During the assessment missions the team members had meetings and interviews with governmental officials at the national (ministries, state committees, national agencies etc.), regional and local levels (including municipal authorities), representatives of private businesses and the banking sector (including project developers and energy service companies), state-owned companies, business associations, academic and research institutions, non-governmental organizations (NGOs) and international organizations.

The National Coordinators and representatives of the National Participating Institutions were responsible for arranging these meetings and interviews in their respective countries. They also provided responses to questionnaire developed and distributed by the Project Management Unit before the assessment missions, and valuable comments and contributions related to their specific countries in the preparation of this report. The latest available energy balances and annual economic data for the countries are presented in the Annex.

Schedule of the assessment missions

The schedule for the assessment missions to each participating country was arranged by the Project Management Unit in cooperation with the National Participating Institutions and consultants. The missions were conducted as follows:

- Albania: 29–31 October 2008
- Belarus: 14–17 October 2008
- Bosnia and Herzegovina: 21–24 October 2008
- Bulgaria: 21–23 October 2008
- Croatia: 27–28 October 2008
- Kazakhstan: 17–21 November 2008
- Republic of Moldova: 4–6 November 2008
- Russian Federation: 27–30 October 2008
- Serbia: 29 September–2 October 2008
- The former Yugoslav Republic of Macedonia: 27–28 October 2008
- Ukraine: 30 September–3 October 2008

The mission to Romania did not take place because the country has temporarily suspended its participation in the project. However, the assessment was done based on the existing data and research available to consultants.

Structure of the Report

Following the recommendations of the Ad Hoc Group of Experts on Energy Efficiency Investments for Climate Change Mitigation, and taking into account the objectives of the assessment missions, the Investor Interest and Capacity Building Needs Report is structured according to the following major areas:

- Financial environment and major barriers to financing energy efficiency and renewable energy projects
- Energy efficiency and renewable energy project development and finance capacities;
- Public and private sector investor interest in the Eastern European Energy Efficiency Investment Fund
- Capacity building needs in the countries of the region for successful development of bankable project proposals in the area of energy efficiency and renewable energy sources
- Energy Efficiency Business Development Course Programme
- Conclusions and recommendations for future project activities.

The Ad Hoc Group of Experts on Energy Efficiency Investments for Climate Change Mitigation at its thirteenth session (2-3 March 2009) approved the conclusions and recommendations set out in this document. The information cut-off date for this publication was 31 March 2009.

Map 1: Map of the countries participating in the Project

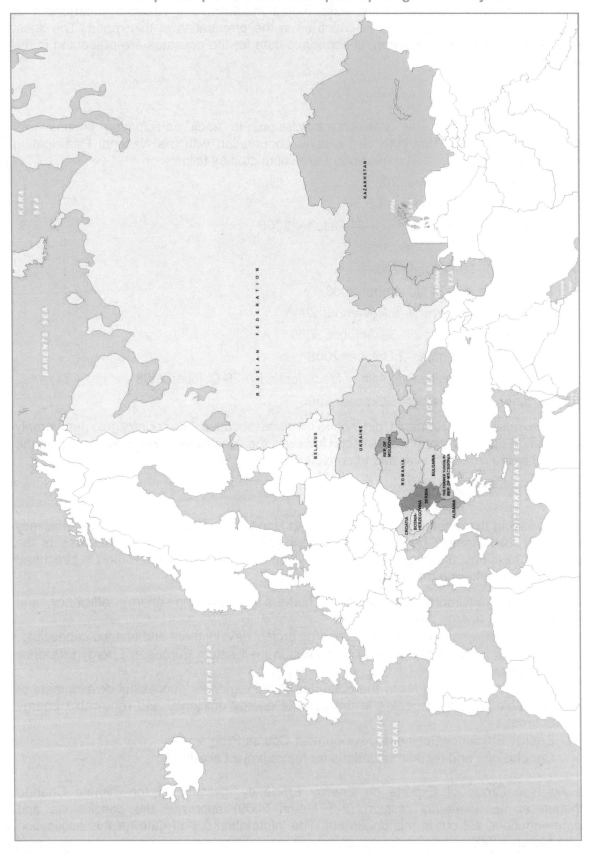

INVESTOR INTEREST AND CAPACITY BUILDING NEEDS

OUTCOME OF THE ASSESSMENT MISSIONS

ALBANIA

Energy overview

Since heating in the building sector is mainly done by electricity, it plays a very important role in the Albanian energy system (about 23.5 per cent of the total primary energy supply). Almost 100 per cent of the electricity produced in Albania is generated by hydropower, but a substantial part of the electricity supply is covered by imported energy (about 48 per cent at present according to the Albanian Power Corporation KESH, the state owned electricity production company). Due to lack of capacity, power cuts in the electricity supply are frequent. (Source: Energy Balance 2007).

Power production (KESH) and transmission (Transmission System Operator (TSO)) remain state owned, while the privatization process of the distribution system was concluded in 2008.

Electricity losses through transmission and distribution, both technical and non-technical, are around 35.7 per cent. One condition in the negotiations on the privatization of the distribution system was to reduce these losses to 16 per cent.

A priority for the Government is to prepare the legal framework and financial environment facilitating private investments also in the energy sector – "everything that could be private should be privatized".

According to the National Power Law, an independent Electricity Regulatory Authority (ERA) has been established to set rules and tariffs. According to ERA, the principle used today is cost +, in future the tariffs will be based on the principle of marginal costs. Electricity prices are still below market value (2008: average 7 euro cents/kWh). However, there has been a price increase and ERA is developing plans to gradually move to full-cost pricing in a way that recognizes social impacts. During the assessment mission the President announced that the electricity tariffs would not be increased in 2009.

There are no district heating systems in operation in Albania, direct electrical heating is the most common, and some buildings and building complexes have their own boilers and water based heating systems (by radiators).

To prepare domestic hot water by solar collectors also seems to be an interesting business for Albania, both to reduce the electricity consumption for this purpose, and to increase the reliability of domestic hot water supply in view of the frequent power cuts that are experienced.

This might be one of the reasons why for the time being in Albania, renewable energy sources (RES) seems to have a higher priority than energy efficiency (EE).

KESH earlier carried out a number of end use energy efficiency projects and awareness raising and education activities aimed at reducing energy demand, and thus possibly reducing the power cuts. As the only electricity provider, they felt responsible for informing end users about what they could do to save energy, which could further allow them to increase the tariffs. After the privatization process, KESH will await further initiatives and follow up on the activities from the new Distribution Company.

Recognizing the importance of the energy sector in supporting overall sustainable development in Albania, the Government has prepared a National Energy Strategy (updated), not yet approved, with the following objectives:

- Establish an efficient energy sector from both a financial and technical point of view
- Establish an effective institutional and regulatory framework
- Increase the security and reliability of energy supply in general and in electricity in particular, at both national and regional levels
- Increase energy efficiency in the generation and use of renewable energy sources aiming to achieve minimal environmental pollution
- Optimize the supply system with energy resources based on the least cost planning principle and minimal environmental pollution
- Complete the restructuring process of energy companies
- Establish a competitive electricity market according to European Union (EU) requirements for electricity sector reforms and Albanian obligations under the Athens Memorandum to support the development of South-East Europe.

Consistency with EU requirements is a key consideration in the development of the Albanian energy sector today.

As part of its programme to achieve the aims of the National Energy Strategy, the National Energy Efficiency Law sets out plans to improve energy efficiency, especially in the industry sector. There are some other legal frameworks which concern the household and services sectors. Specifically there are laws on "energy saving in buildings" and "energy code".

Reducing transmission and distribution system losses from their current levels to 6 per cent by 2020 is a high priority, effectively delivering around a 20 per cent increase in available electricity.

However, the Energy Efficiency Law is not yet implemented. For instance, the building codes, requiring better insulation, and installation of water based heating systems (radiators with distribution pipes) in buildings are implemented to a very low degree even in public buildings.

The National Energy Strategy (updated) is in the final stages of approval, and Energy Efficiency Law is to be updated. In the updated National Energy Strategy the diversification of electricity production is given high priority.

A Renewable Energy Law is being developed, the first draft of which is be presented shortly.

The decentralization process is ongoing, and municipalities are getting budget responsibilities for operating public buildings, street lighting and water supply. These responsibilities are managed by a department or unit (person) responsible for "infrastructure". Hospitals are still partly funded by the state budget. Municipalities have recently been given the right to obtain loans. The Albanian Association for Municipalities provides the municipalities with selected information on various subjects, and energy efficiency is one of the hot topics.

The Ministry of Environment is the national designated authority (DNA) for Clean Development Mechanism (CDM) projects in Albania. However, all climate change mitigation activities are dealt with by the local office of the United Nations Development Programme (UNDP).

National priority areas for EE and RES

Based on the National Energy Strategy, some measures are foreseen, including in the performance of KESH, to improve the existing legal framework, etc.

To improve the performance of KESH, there following can be foreseen:
- reduction of electricity losses in distribution and transmission and raising billed electricity consumption;
- installation of meters where they are lacking;
- elimination of tariffs related to cross-subsidies for all public, non-public and private consumers.

Efforts will continue to focus on the legal approximation of the energy sector to that of the European Union countries through the adoption of the *acquis communautaire*. To this effect:
- The law on renewable energies has already been drafted;
- The existing energy efficiency law will be improved.

Some energy efficiency measurements are foreseen in different economic sectors:
- Using energy saving bulbs and other more efficient electrical appliances;
- Thermo isolation in service and household sectors;
- Public transport;
- Orientation towards alternative energy resources;
- New high efficiency technologies in industry, etc.

The use of renewable energies is another government priority and is reflected even in this strategy. Some of the projects to be implemented in this field are:
- Through bank credits of small hydro power plants (HPPs)
- Preparation of a strategic plan for the development of hydro-energy resources through studies in this field
- Increase in the penetration of solar panels for hot water in the residential and service sectors.

Interest in receiving equity and mezzanine financing

Some companies have expressed a potential interest in receiving equity or mezzanine financing from the EE Investment Fund.

The Albanian Investment Fund expected that there would be a substantial interest in the market for the EE Investment Fund.

According to the Albanian Bank Association, the EE Investment Fund could be highly instrumental in getting RES projects carried out. Once the first positive experience is documented, it would be easier to realize a large number of projects.

KESH is interested in establishing new production facilities with other investors, including the EE Investment Fund.

Small Hydro-Power Plant in Albania

Barriers to financing EE and RES

There are several barriers to increased energy efficiency and utilization of renewable energy sources in Albania:

- Lack of capacities and skills in how to prepare project proposals properly;
- Low overall electricity price levels;
- Lack of public awareness;
- Lack of sufficient capacities to deal with energy efficiency in municipalities;
- Lack of equity for RES projects;
- Purchase guarantee only for electricity from small hydropower (SHP) <15 megawatts (MW);
- Legal framework in view of European Union directives.

Due to the absence of operating district heating systems, the low level of heating of public (and private) buildings, low energy tariffs and insufficient public priority and awareness, the market for energy efficiency investments is limited.

Because of the lack of funds, hospitals, schools and kindergartens are not properly heated, resulting in low indoor temperatures during the heating season. It will thus be difficult to achieve real energy savings, which could be a major barrier to ESCO (Energy Service Company) business. Furthermore, there seem to be limited incentives for energy savings in budget funded objects.

According to the World Bank, ESCOs cannot compete on the production side due to the way the Albanian energy market is organized. On the demand side, the tariffs are too low for an ESCO to do business. With substantially increased tariffs and secured public budgets, ESCO business could be considered. In households, measures reducing consumption above 300 kWh/months could be profitable due to the relatively higher tariff for this level (10 euro cents/kWh).

Again according to the World Bank, the main problem regarding RES projects is lack of equity. Since the purchase guarantee only is for electricity from SHP (<15 MW), the projects are small, they need to be evaluated individually, and the transaction costs are high. Higher tariffs are needed to make the projects profitable enough.

Incentives

A priority for the Government is to prepare the legal framework and financial environment facilitating private investments also in the energy sector – "everything that could be private should be privatized". The Ministry of Finance is looking at how best to benefit from the various international initiatives on RES, where hydropower projects should be prioritized. According to the Ministry, even large hydropower plants could be privatized in the future.

A model for calculation of feed-in tariffs has been established, and KESH is obliged to buy electricity produced by small hydropower plans (<15 MW). The main principle of the model is average import prices + 10 per cent, recalculated annually. For 2008 the feed-in tariff was approximately 8.1 euro cents/kWh. The limit of 15 MW might be a barrier to foreign investments.

No such regime is established for electricity produced by wind or solar energy photovoltaic (PV).

The Ministry of Economy, Trade and Energy has established the Albanian Investment Fund to be a contact point and coordinator for Foreign Direct Investments in Albania. This Investment Fund (not having any funds for investments) is working closely with the Chamber of Commerce and all their member companies.

Financing schemes

According to the Ministry of Finance, there are no plans today to capitalize the proposed Albanian Energy Efficiency Fund and the Environmental Fund.

The Italian Government has provided 5 million € to a EE & RES Fund within the Ministry of Economy, Trade and Energy providing 30 per cent grant to projects with high reductions of CO_2.

The KfW Bankengruppe's Energy Sector Programme promoting energy efficiency and renewables, has the objective of contributing to improving energy supply and protecting the climate and environment by promoting the environmentally compatible and efficient generation and use of energy. Project design: partial loan guarantees are provided for local commercial banks through a renewable energies facility to create the conditions for investments in small hydropower plants (guarantee component). Moreover, investors are supported in the preparation of individual projects and the banks participating in the programme are assisted in conducting loan appraisals and in the selection of programmes to be financed (advisory component). Projects to increase efficiency in energy end use are also supported of 9 million € in FC funds, 2 million € are to be spent on advisory services, and the remaining 7 million € will be used as a pool to fund accompanying guarantees and energy efficiency measures (50/50 for EE and RES).

In the Italian Development Programme for small and medium-sized enterprises (SME) (25 million €), projects from the energy sector are also eligible (50 000 - 500.000 € per project).

The Government of Japan has expressed interest in providing grants through the Albanian Government to support the development of SHP.

When the privatization of the Electricity Distribution Company is concluded, the World Bank would do a project on providing energy efficient lamps to primarily households (the distribution company has about 1 million customers, including 800.000 households).

Various UNDP projects:
- Building capacity to access carbon finance in Albania (June 2007 - June 2009)
- Solar water heating (that foresees the installation of 70,000 m^2 of solar panels in Albania).

European Bank for Reconstruction and Development (EBRD):
- The Western Balkans Local Enterprise Facility (LEF) was established in 2006 as a €32 million investment channel, to which Italy initially contributed €12 million. The aim is to provide mainly loan and equity financing to promising SMEs in the region
- Western Balkans SME Finance Facility. Provides loans to local banks and leasing companies in the region, with donor support from the European Agency for Reconstruction and Italy
- Western Balkans Sustainable Energy Direct Financing Facility. €50 million to provide direct individual loans of between €1 and €6 million to industrial energy efficiency and renewable energy projects. The facility is complemented with grant funding for technical assistance for project identification, preparation and implementation verification.

The United States Agency for International Development (USAID) Local Governance Programme in Albania works with ten municipalities throughout Albania to foster local economic growth, improve local governance, and strengthen civic and private sector engagement in local development.

ESCOs

There is no existing or planned ESCO.

Banking sector

Because of the 1997 crises in Albania, the Government and the banking system became more conscious of the necessity of long term financial strategies and now there are state funds available to cover five months of imports.

According to the Albanian Central Bank, the economic situation in Albania is quite stable. Some key figures (2007):
- 6 per cent annual growth of GDP
- $ 3.674 GDP per capita (Source: Ministry of Finance)
- 2.9 per cent inflation
- 13,2 per cent official unemployment rate (the real rate assumed to be twice this figure)
- 3.4 per cent budget deficit (per cent of GDP)
- 6.25 per cent interest from the Central Bank to the 17 commercial banks
- 5-6 per cent annual depreciation of the local currency lek
- in average 17 per cent equity capital in the commercial banks (min 12 per cent required)

The main recommendation from the Central Bank now is consolidation. There has been a rapid increase in providing loans over the last few years, and some crediting restrictions are considered.

The Central Bank is not involved in project financing, but has the impression that investments in RES are promising. However, there are some challenges related, for instance, to property rights.

According to the Central Bank, the legislation for utilizing equity funds has been in place for 10 years, but there has been no experience – "it would be very interesting to develop this type of financial scheme for Albania".

According to the Albanian Bank Association, commercial banks have no experience in project financing. Some of the banks are in discussions regarding lending to SHP projects, but are reluctant to enter into this business, being used to having physical collateral as a guarantee. SHP also involves relative small investments (on average €1 million for 1 MW capacity, feed-in tariffs only guaranteed for SHP with a total capacity of less than 15 MW).

The National Commercial Bank has been involved in evaluating some SHP projects. Its experience and comments include:
- There is a great potential for SHP in Albania
- There is an increasing interest and willingness to facilitate project realization from the Government
- There is an increasing interest from companies (project owners), but lack of knowledge
- The guarantee issue is difficult, and the purchasing agreement is not clear
- The project development period is long, including authority approval up to license (based on pre-feasibility study), and then preparing bankable feasibility study
- Due to lack of experience, the completion risk (final costs and construction period) is evaluated to be considerable
- Long construction period before cash is generated to start repaying loans
- The equity from the project owner (lender) is mainly land, sunk costs, etc.
- Additional cash is needed to enable project realization.

The National Commercial Bank therefore believes that the Eastern Europe Energy Efficiency Investment Fund could be highly instrumental in getting such projects implemented. Once the first good experience is documented, it would be easier to carry out a large number of projects.

Brief conclusions and recommendations

- The legal and regulatory framework for business development in Albania is developing rapidly and seems to be beneficial; the Government seems to be very supportive of ensuring foreign investments and private initiatives
- Lack of equity seems be a main barrier to investments in the energy sector
- Equity participation in the SHP sector seems to be the most promising, public purchase commitments for electricity only from power plants of less than 15 MW might be a barrier
- Equity participation in wind farms might also be promising
- The market does not seem to be ready for a Fund involvement in an ESCO establishment.

Energy efficiency and renewable energy sources project development and finance capacities

Existing and prospective EE and RES projects

Concessions have been given to a number of SHP Plants (upgrade and/or building new), and national and international companies and investors have been invited to give offers. Interest has been shown, and feasibility studies have been prepared, but no contracts for project realization have yet been signed. A new 97 MW, € 100 million thermal power station in Vlora is planned to be completed by spring/summer 2009 (owned by KESH, 30 per cent equity, 70 per cent loan by World Bank, EBRD and European Investment Bank (EIB).

KfW: 400-kV Transmission Line Montenegro – Albania. The project aims to expand the transmission capacity for the cross-border high-voltage line between Tirana (Albania) and Podgorica (Montenegro) by adding a 400-kV overhead line to the existing 220-kV line. The necessary investments will cost about €54 million.

The scope of the project has been reduced because the Italian Government will finance the section between Elbasan and the planned substation near Tirana. A loan in the amount of € 43.9 million was provided to finance th e remaining section to the Albanian-Montenegrin border and to expand the substations both on the Albanian and the Montenegrin sides. The project is expected to be completed by mid-2009.

Electricity from SHP and wind-farms seems to have the best prospects in Albania today. There are ongoing discussions regarding establishing a sea cable to export wind-produced electricity with Green Certificates to Italy. If there were a proper grid allowing export to Italy, Austria, Switzerland, etc., the profitability of SHP projects would be even better.

Regarding energy efficiency, there is some interest in the private sector (service and tourism). There is very low activity in existing industry, and new industry is expected to be constructed in a more energy efficient way.

Assessment of investment project development skills

Based on the meetings arranged by the NC and NPI, it is the impression of the consultant that the level of capacities and skills operationally available for the development and financing of energy efficiency and renewable energy investments in Albania is low. Therefore, there is a need for capacity building.

Public sector

The Ministry of Economy, Trade and Energy and the National Agency for Natural Resources confirmed that project development capacity building is greatly needed. Lack of public awareness is another challenge.

Private sector

There is no existing or planned ESCO, and therefore no experience with this type of operation.

Regarding energy efficiency in public and private buildings, the very few projects were demonstrated during the assessment missions. Without a number of projects being realized, the capacities and skills on development of bankable energy efficiency projects is most likely not there.

Further capacity building seems to be beneficial also for the NGOs involved in the area of work.

Assessment of equity and mezzanine financing business development skills

The feasibility studies were prepared for a number of SHP projects, but the estimation of the quality of this study was not done. There is no evidence of implementing any RES project (as well as any decision to do so), thus this could also indicate that the capacities and skills available are limited.

The establishment of any special purpose companies was also not evident, and therefore, the experience, knowledge and skills on equity and mezzanine finance are to be very limited. This was explicitly expressed by the representative of the Central Bank.

Brief conclusions and recommendations

- There is a substantial need for capacity building in Albania, both related to investment project development and equity and mezzanine finance business development
- Capacity building is also needed to strengthen the various institutions involved in the field of EE and RES, and better coordination and cooperation between them would be helpful.

Investor interest

The issue of potential participation in the Eastern Europe Energy Efficiency Investment Fund was presented and discussed with several government counterparts and representatives of private sector during the mission. Below is a brief summary of the results.

Public sector

The Ministry of Economy, Trade and Energy is "in principle interested, but need to study a concrete proposal".

The Ministry of Finance has "no capacity today to invest in the fund, but this could be evaluated in the future if there were several Albanian projects of interest for the fund."

Private sector

KfW could be interested in participating in the Fund, this would need to be discussed with the Headquarter in Frankfurt.

The Central Bank of Albania does not expect the commercial banks to show interest in investing in the Fund.

The interest from international financial Institutions needs to be clarified by their headquarters, and not through the branch offices in Albania.

Potential partners for project co-financing

The Albanian Bank Association stated that they expected several of their member banks to be interested in co-financing projects with the Fund.

KfW could also provide loans to projects where the Fund is involved with equity.

Brief conclusions and recommendations

Real interest from public and private investors in Albania to participate in the European Energy Efficiency Investment Fund seems to be limited. The interest from international financial Institutions needs to be clarified by their headquarters, and not through the branch offices in Albania.

BELARUS

Energy overview

Existing energy resources, energy dependence for primary and secondary energy resources, production of electricity and of heat, use of renewable energy sources (RES)

Electricity sector

The own resources (including renewables) account for 17-18 per cent of the country demand and include small amounts of crude oil, biomass (mainly wood), peat, coal, natural gas and hydropower. Most of the demand is covered by imports of gas, oil, coal and electricity from the Russian Federation. Most of the imported crude oil is refined in Belarus for export and domestic consumption. The imported natural gas represents 61 per cent of the total energy consumption in the country. 72 per cent of total natural gas consumption is used for production of heat and electricity and 28 per cent for technological needs.

The main producer of electricity is the State Industrial Corporation (GPO) Belenergo.

To ensure the energy security of Belarus, the possibilities for diversification of the supply of gas and oil, maximum use of local resources, development of renewables and increase of energy efficiency are considered.

Lepel Small Hydropower Plant (Vitebsk Oblast)

The main energy exports include: petrochemicals, lubricants, electricity and peat. The amount of natural gas from the Russian Federation transited through Belarus is the second largest after Ukraine. The gas transport network is one of the most important strategic infrastructure objects in the country. The volume of transit gas was 49.5 billion m^3 in 2007. The capacity of the network is 63 billion m^3 per year.

The total installed electrical power in the country is around 8,000 MW, of which 44.2 per cent is large condensing thermal power plants, 52.7 per cent is combined heat and power plants and the rest is small thermal and hydropower plants. The installed capacity is higher than the country's domestic demand. The demand in 2007 was about 36 billion kWh. It is forecast that the demand in 2020 will be 43-50 billion kWh and the heat demand will be 81-88 million gigacalories (Gcal). To meet demand it is planned to build an additional 1,500 MW of generating capacity by 2011 and an additional 4,000 MW by 2020. There are plans to put into operation the first unit of a nuclear power plant in 2016. Electricity generation for export purposes is planned.

The high-voltage grid of the Belarusian electricity system is part of a large high-voltage ring covering also the Baltic countries and the Russian Federation. The grid is connected to Russia, Lithuania, Poland and Ukraine. GPO Belenergo supplies around half of the heat power in Belarus. The other half is supplied by local heating systems, which belong to municipalities or industrial enterprises.

The fuel and energy complex of Belarus is overseen by the Ministry of Energy. The commercial activities are performed by legally independent but fully state-owned enterprises: GPO Belenergo, GPO Beltopgas and OJSC Beltransgas. In the system of the Ministry of Energy there are design, research, installation and maintenance complexes, which have well educated professionals and equipment to complete and maintain the full cycle of the necessary works in the energy sector. At present, state-owned and private companies are introducing small local sources of heat and electricity.

Structure of electricity and heating tariffs

The tariffs for residential users are set by the Council of Ministers, and for industrial users by the Ministry of Energy. The Ministry of Economy performs the role of "independent" regulatory body and regulates the tariff policy through its orders on prices and tariffs for energy for end users. All orders undergo judicial expertise in the Ministry of Justice.

The tariffs for locally generated and sold heat energy are coordinated by the local municipal bodies.

The rates for electricity and heating depend on the type of consumer and the use of the building, the time and period of use (day/night and heating/non-heating season) and other factors.

The price of electricity is set by the type of consumer: industrial consumers with installed consumption capacity of 750 kilovolt-amperes (kVA) or more (dual tariff); industrial consumers with installed consumption capacity up to 750 kVA (single tariff); transport; non-industrial users, including budget organizations; hospitals; households; agriculture.

The prices for heat energy produced at GPO Belenergo depend on the type of use and the parameters of the heat carrier. It differs for different consumers (households, budget organizations, industrial companies, etc) and different regions. The price of the heat energy produced by the Minskcommunteploset (Minsk communal heat network) is different for households and other consumers.

For households, the tariffs for electricity and heat at the end of 2008 were respectively: 145 Belarusian roubles (Rbl) per kWh and Rbl 37,580.70 per Gcal and were increased at the beginning of 2009 to Rbl 173.3 per kWh and Rbl 43,458.30 per Gcal (exchange rate Rbl 2,200 for $1).

The payment system is undergoing modification through the introduction of new procedures and measuring equipment, based on modern electronics and automated control and measurement systems. These measures are being implemented in accordance with the Resolutions of the Council of Ministers on measures for introduction of automated system for control and accounting of electricity (2005) and the Order of the Ministry of Energy on the concept of measurement of electricity (2005).

There are plans to replace preferential and fixed national tariffs by a set of differentiated tariffs to be applied to all economic sectors.

Level of priority given to EE and RES in country's energy policy

Existing legislation and regulations mandate the municipal and governmental bodies to support the enterprises in privatization and leasing assets in the areas of alternative energy and small energy generation. It is compulsory to connect these objects to the grid, irrespective of their ownership, as well as to pay for the generated energy.

The Order of the Ministry of Economy (2006) provides certain preferences and incentives for the generation of electricity from alternative and renewable sources of energy and for the small power plants.

The main task is to attract private and foreign capital to finance energy efficiency and renewable projects. To achieve that it is necessary to create a conducive institutional and economic climate.

Investments in the energy sector

Implementation of the energy policy of Belarus (see sub-section *Legal, regulatory and policy framework* below) requires significant amounts of investments. The requirements for renovation and development of the energy system of Belarus according to various programmes in the energy sector are estimated at over $3.1 billion up to 2011. About half of all investments are planned to be directed to co-generation facilities and about a quarter to the development of power grids.

The structure of capital investments in the renovation of the energy system by the sources of financing is planned as follows (rounded):
- Innovation Fund of the Ministry of Energy – $356 million
- Amortization – $1,123 million
- Debt capital – $452 million
- Budgetary provisions – $923 million
- Companies profits – $248 million

The share of budgetary allocations and debt capital are expected to grow considerably in the coming years and reach 30 per cent and 15 per cent respectively.

The following volumes of capital investments were developed in 2008:
- By GPO Belenergo – $ 643 million
- By GPO Beltopgas – $ 157 million
- By OJSC Beltransgas – $ 125 million

Energy saving and renewable energy are also expected to receive considerable financing. Total capital investments for these purposes in all economic sectors, excluding investments of GPO Belenergo, for 2006-2010 are envisaged in the range of $ 5,200 – 5,850

million. The structure of the capital investments in energy saving and renewable energy by the sources of financing (as a share of the total) is planned as follows:
- Financial resources of companies – about 46 per cent
- Innovation funds of the ministries – about 26 per cent
- Budgetary provisions – about 15 per cent (of which about 2 per cent from the national budget, and about13 per cent from municipal budgets)
- Borrowing, debt capital – about 13 per cent.

Annual capital requirements (excluding fuel and energy complex) for implementing energy saving and renewable energy projects are estimated as follows (rounded): 2006 – $ 600 million; 2007 – $ 863 million; 2008 – over $ 960 million; 2009 – over $ 1260 million; 2010 – over $ 1606 million.

Financial environment in energy efficiency and renewable energy

Legal, regulatory and policy framework

The following Government institutions are involved in developing and implementing policies and actions in the fields of energy, energy efficiency and renewable energy sources:
- Ministry of Energy, GPO Belenergo, GPO Beltopgas, OJSC Beltransgas
- Department for Energy Efficiency of the State Committee for Standardization
- Ministry of Natural Resources and Environmental Protection

The Ministry of Natural Resources and Environmental Protection is also responsible for the policies and activities in the area of climate change.

The major legislation and policy documents (strategies, programmes, action plans) in the areas of energy, EE and RES are:
- Law on energy savings (1998)
- Concept of energy security of Belarus (2007)
- Directive of the President "Economy and saving – main factors of economic security of the State" (2007) and Resolution of the Council of Ministers on measures for its implementation
- State programme for energy saving 2006-2010
- Governmental programme for the modernization of productive assets of the energy system of Belarus, energy savings and increase of the use of own energy resources for the period up to 2011
- Special purpose programme for providing not less than 25 per cent of the electricity and heat from local fuel and use of alternative sources of energy for the period up to 2012

Sectoral ministries develop specific programmes, including short- and medium-term plans for energy savings in their respective industries.

There are numerous norms, standards and guidelines on the efficient use of fuel and energy. Several pieces of legislation are under development, including a draft law on non-traditional (alternative) and renewable sources of energy and a draft law on electricity. According to statements from Government officials, Belarus plans to adopt a new energy law in 2009 allowing privately owned energy companies. Several policy documents, such as a draft programme for the development of wind energy and a draft programme for the development of small hydropower plants, are also under preparation. A draft programme for the development of wind energy is under consideration by the Government at the Prime Minister level (as of February 2009).

The development of technical standards and norms is among the priorities of the Belarus policy in EE and RES. A *Programme for the development of a system of technical regulations, standardization and compliance assessment in the energy saving field* has been drafted. It contains a number of organizational, technical, economic and production benchmarks and envisages the development of 129 technical regulations. The Programme also sets the requirement for state standards to comply with the relevant international and European standards and EU Directives.

National priority areas for EE and RES

Belarus has a relatively low potential for solar, geothermal, wind and hydropower. The most significant renewable energy resource is wood and other biomass sources (about 1,000 MW_e technical potential). Over 400 boiler plants and mini-combined heat and power (CHP) plants are operating in the country using the combined combustion of wood and other fuel such as coal. Currently, a large range of water-heating boilers and mini-CHP (with capacity from 60 to 5000 kW) operating by combustion of wood and timber waste is produced in Belarus.

The priority renewable projects in Belarus are mostly hydropower, including:
- Polotsk – 23 MW, $109 million
- Verkhnyaia Dvina – 29 MW, $138 million
- Beshenkovichi – 30.5 MW, $145 million
- Vitebsk – 50 MW, $230 million

In addition, there are several planned projects (Orsha – 4.9 MW, Shklov – 5.5 MW, Zhlobin – 9 MW, Vilyakovka – 11 MW, Mogilev – 15 MW, and Rechitsa – 24MW).

Several small wind projects are under development.

A bigger potential for savings can be found in the rehabilitation of the conventional power plants and construction of new more efficient gas power plants. There is also a significant programme for construction and rehabilitation of transmission lines and substations.

It has to be noted that with the current tariff structure, the projects listed above offer rather poor returns on investment: the simple payback period is from 7 to 15 years and the corresponding internal rates of return are about 1 to 5 per cent.

Interest in receiving equity and mezzanine financing

The Government of Belarus claims that attracting investments, including foreign direct investments (FDI), as one of its priorities. The Government has established a special agency to attract and manage investments in Belarus. There is also considerable interest in attracting investments specifically in the energy sector, particularly in the area of EE and RES. Energy efficiency and energy security are considered high priority by the Government.

However, there is little understanding of the financial realities among the officials at the ministries and the state-owned companies. An important feature of the Belarusian economy is that the majority (over 75 per cent) is state-owned. There is little understanding of the concept of mezzanine debt or similar instruments. The idea that an investor may expect 15 per cent return on an investment was not well received by companies and officials interviewed during the assessment mission. This issue may require additional research.

Barriers to financing EE and RES

There are a number of issues affecting the interest, capacity, willingness and possibility to invest in renewable energy and energy efficiency projects in the current financial climate.

- The legal and fiscal mechanisms for converting savings from EE measures into revenue, which can be used to service loans, are not yet fully developed, despite the existence of respective framework legislation.
- Borrowing capacity of municipalities is relatively low.
- In the current economic situation, the Government may be less inclined to support subsidies for renewable energy and to support EE projects on a large scale. Availability of public spending may become more restricted. However so far the Government has not announced any intention to decrease its support for renewable energy and EE projects, which is encouraging.

Incentives

There are a number of legal norms and acts to stimulate the activities of enterprises to reduce the consumption of fuel and energy and implement energy saving technologies. These include preferential loans, State subsidies, the use of saving to pay bonuses, systems of penalties and sanctions for non-rational use of energy resources, etc.

Financing schemes

Sources of financing in the field of energy efficiency: own funds of enterprises; innovation fund of the Ministry of Energy for energy efficiency and other innovation funds; national and local budgets; loans. The main source of finance for energy efficiency projects is the own funds of the enterprises. The State support is in equity participation in the form of local and national budget investments and governmental innovation funds. It has to be emphasized that the majority of companies in Belarus are state-owned.

For the development of local and alternative sources of energy the following financial sources are considered: own funds of enterprises; funds of the sectoral innovation funds; funds from the innovation fund of the Ministry of Energy; loans; funds from national and local budgets for the modernization of state-owned enterprises; governmental funding for R&D in the field of use of local fuels and alternative sources.

A number of projects are being implemented using assistance (both grants and loans) from the EU, World Bank, EBRD, UNDP, and Global Environment Facility (GEF).

ESCOs

Several ESCOs operate in Belarus, including BelinvestESCO, Vneshenergoservice and Gmmotory. They are implementing projects under energy performance contracting schemes.

Energy efficiency and renewable energy resources project development and finance capacities

Existing and prospective EE and RES projects

Belarus ratified the Kyoto Protocol to UNFCCC in 2006 and is an Annex I party to the Convention. However, Belarus cannot currently trade emissions or participate in Joint Implementation (JI) projects, as the amendment to the Protocol to allow that needs to be ratified by 75 per cent of the parties to the Protocol (by the end of 2008 only 5 of the 175 countries had ratified it). If Belarus overcomes this legal hurdle there is a significant identified potential for implementation of JI projects.

The necessary legislative acts and regulations are being adopted and national registry and national implementation institutions established.

Belarus actively prepares JI projects. A pipeline of projects has been identified. The Ministry of Natural Resources and Environmental Protection has done an expertise of Project Idea Notes for small heat and power plants that use wood, cogeneration and biogas plants, small hydro, waste utilisation, buildings efficiency. Currently a number of project design documents are under preparation. Among the advantages of the Belarusian JI projects are:

- Large portfolio
- State guarantees for implementation
- Defined national procedure for approval
- Possibility for portfolio diversification by the purchaser

Currently, some projects are being implemented under voluntary emission trading schemes.

Most of the EE and RES projects are implemented by state-owned companies with governmental support. However, there are several private companies that implement such projects, among them the joint stock companies Naftan, Polimir, Belsol, and Belsolod.

Assessment of investment project development skills

Public sector

Policy- and decision-makers at the national and local levels would benefit from training in financial engineering and business planning basics, which provide clear and concise information on what the general requirements are of the international financial institutions for an investment project. This would provide understanding and support at the decision-making level for the identification, selection and development of energy efficiency and renewable energy investment projects and the preparation of bankable project proposals. Large state-owned companies have experience with developing project proposals, however they are mostly intended for local banks (also mostly state-owned or with the state being a majority shareholder) whose decisions on loans may not always be taken based on purely business considerations.

Private sector

The private sector in Belarus is represented mostly by small and medium-sized enterprises. Such companies will benefit from detailed training and assistance in preparation and presentation of their business plans, with particular emphasis on documentation completeness, financial projections and modelling.

Assessment of equity and mezzanine financing business development skills

The development of skills related to the use of mezzanine debt is closely related to the availability of knowledge and skills to prepare and present a viable business plan with an addition – the use of mezzanine debt to close any gaps in the project's financial plan. There is little knowledge of the use of subordinated or mezzanine debt. General training on business planning can be accompanied by the provision of information on such instruments.

Investor interest

The interviews with the potential investors and financial institutions reveal that at present the lending to new clients has practically ceased, but banks like EBRD, International Finance Corporation (IFC) and World Bank are in the process of preparing of new financial facilities.

The provision of mezzanine debt is closely related to the availability of senior debt. Should the planned facilities of the international banks materialize next year, there will be demand for mezzanine debt.

Public sector

There was no clear expression of interest from the Government officials interviewed during the assessment mission for Belarus to become an investor in the Fund. However, taking into account the goals of existing governmental innovation funds and investment programmes, there is a potential for co-financing of the EE and RES projects by these funds.

Private sector

The private sector in Belarus does not play a major role in the national economy, and there are no expectations for private companies from Belarus to become investors in the Investment Fund. However a number of Belarusian banks, such as Belinvestbank, Belarusbank and UniCredit Bank (Belarus) confirmed that they may be interested in participating in some form of co-financing of projects with the Fund, through some of the Government supported schemes, or in separate projects.

Potential partners for project co-financing

In addition to the co-financing opportunities described above, several international financial institutions (IFIs) (for example EBRD, EIB, and the World Bank) may become partners in co-financing specific projects of the Investment Fund.

Brief conclusions and recommendations

Despite the declared interest in investments, there are a number of barriers facing any potential investor and a perception of risk, which makes investments in this sector in Belarus difficult. Some of the main issues include:
- Subsidized domestic tariffs for electricity and heat;
- Difficulties in monetizing the realized savings – for example, savings achieved in an organization that receives funding for its operations from the state budget will not be available to it and remain in the state budget, thus they cannot be used for repayment of the investment.
- Investing in projects owned by State and municipal entities (in Belarus this represents the vast majority of all projects) always raises the issue of guarantees and collateral – can the project be guaranteed by the State (this is usually a slow and cumbersome process), what is the value of such guarantee, can municipal or state property be accepted as a collateral and many other similar legal and financial issues.

The problems described above lead to a preliminary conclusion that an equity investment in State and municipal owned entities is difficult. Additional research on the issue is necessary.

An equity investment in a particular project, where a Special Purpose Vehicle is established and the State is one of the shareholders, may be possible, but particular details of such an investment need to be explored.

Belarus has plans to adopt a new energy law allowing private ownership of energy companies. However before the law is adopted and becomes operational it is difficult to judge what implications it may have for EE and RES investments.

Interest expressed by several banks in Belarus in co-financing projects of the Investment Fund is encouraging.

BOSNIA AND HERZEGOVINA

Energy overview

Bosnia and Herzegovina imports gas, oil and oil derivatives either as raw materials or in processed forms. Its refinery capacity is located in Bosanski Brod. Natural gas is imported by pipeline from the Russian Federation via Ukraine, Hungary and Serbia.

The coal sector is a very important part of the energy sector and of the economy of Bosnia and Herzegovina. It accounts for approximately 50 per cent of the country's primary energy supply. Coal is also the main energy source for electricity production. The coal produced in Bosnia and Herzegovina is brown coal and lignite. The brown coal is of relatively good quality but has sizable quantities of sulphur, ash and humidity. The lignite is also a pollutant variety of coal. About 85 per cent of the coal produced is used for power generation.

There are three power utilities covering three geographic areas of the country as well as a state wide transmission company (Transco) and an independent system operator. The three Electroprivreda (EP) are the Electricity Company of Bosnia and Herzegovina (EPBIH), Electricity Company of Herzeg-Bosnia (EPHZHB) and Electricity Company of Republika Srpska (EPRS). Each EP has its own generation, distribution and retail facilities. Transco is jointly owned by the two entities (Federation of Bosnia and Herzegovina and Republika Srpska), and is responsible for operation and maintenance of the transmission system while the independent service operator operates the system in terms of dispatch and cross border trading.

Total power production in 2007 amounted to 12,175 GWh. The power sector is predominantly based on national resources of coal and hydropower. Power production is one of the strong points in the development of Bosnia and Herzegovina since the country has a surplus of generating capacity. Untapped hydropower resources and coal reserves make the power sector a potentially growing export earner.

Bosnia and Herzegovina is in a relatively strong position with respect to electricity resources and production. Electricity generation from hydro sources is considerable. Depending on hydrological conditions, up to 45 per cent (estimated 6000 GWh in 2010) of total generation may come from hydroplants with installed hydro capacity close to 57 per cent (2000 MW)[2] of total installed capacity. These figures are indicative and depend on hydrology and the year under discussion.

Future expansion of the hydro system can be expected as not all hydro sites have been developed to full capacity. It is estimated that around 40 per cent of exploitable hydro is now in service.   The potential for small hydro development is also significant. The study by Energetski institut Hrvoje Požar (EIHP) identified close to 300 MW of planned small hydro; future potential may be as high as 1000 MW. Other potential renewable energy sources in Bosnia and Herzegovina include biomass, wind, solar and geothermal as well as the small hydro.

The country has strong links to neighbouring countries as its power system was developed to supply power to other parts of the former Yugoslavia. However, currently most thermal power plants are fairly inefficient and strongly need modernization.

There are district heating systems in the urban areas, especially in Sarajevo, Banja Luka, Zenica and Tuzla. The district heating system in Tuzla derives its heat from the Tuzla power plant, being the only example of an operational co-generation installation for domestic heating in Bosnia and Herzegovina. District heating is supplied by various fuels including

---

[2] Energy Sector Study in Bosnia and Herzegovina, final report, EIHP, 2008.

heavy fuel oil (mazut), coal, wood waste in heat-only boilers. The Sarajevo district heating system uses natural gas.

The EU directives apply in Bosnia and Herzegovina through the Energy Community Treaty. To date, there is no state-wide energy strategy for Bosnia and Herzegovina; consequently the country has no existing target level for renewable sources in gross electricity consumption. However, due to its indigenous hydro, the country should have little difficulty in meeting or exceeding current EC targets.

Utilizing the Intelligent Energy-Europe Programme, which aims at encouraging the wider uptake of renewable energies, Bosnia and Herzegovina should be able to increase the level of investment in new and best performing technologies.

Other potential renewable energy sources in Bosnia and Herzegovina include biomass, wind, solar and geothermal as well as small scale hydro. Biomass in the form of fuel wood is in significant use especially in the household sector, with total consumption estimated at 1.5 million tonnes. Estimates of fuel wood use are up to 60 per cent of household heating requirements in non-urban areas. Wind energy potential has been evaluated for a number of sites with areas closer to the Croatian border being the more promising of those evaluated. The total potential at evaluated locations is estimated at 900 MW[3], with the upper limit as high as 2000 MW. Investigations into geothermal potential to date indicate little potential for electricity production, though some use could be made for space heating and recreational use.

Returning to biomass, other than fuel wood use in households, wood industry waste and agricultural biomass offer possibilities either as a fuel or as feedstock for biofuels. Bosnia and Herzegovina has a well developed wood industry with over 1600 sawmills in operation. The annual cut is estimated at 7.4 million m^3 with over 30 per cent becoming waste, either left on the forest floor or as waste in the wood industry. Agricultural waste is the other potentially significant source of biomass, though this is relatively small in comparison with the wood waste. Traditional use of biomass is as a fuel for space heat and hot water and for electricity generation in cogeneration plants. With increased interest in biofuels, increasing attention will be paid to using biomass for conversion into biofuels.

It should be noted that the assessment mission team had meetings on a state level and on the level of Republika Srpska only.

Financial environment in energy efficiency and renewable energy

Legal, regulatory and policy framework

Bosnia and Herzegovina signed in June 1995 the Energy Charter Treaty and ratified Energy Charter Protocol on Energy Efficiency and Related Environmental Aspects in January 2001.

The Law on electric power transmission, system regulator and operator in Bosnia and Herzegovina was adopted in 2002. It was developed to modernize the structure of the power sector, which has been implemented with the creation of Transco and the independent system operator and is ongoing with the unbundling of the EP generation and distribution functions. The legislation does not specifically address issues of energy efficiency or environment.

Considerable progress has been made with respect to the reform and restructuring of the electricity sector. While progress has not always been as rapid as planned it is worth recalling that the sector now is quite different from that of five years ago.

---

[3] EIHP Study, module 12.

A number of laws are in place and as a result Bosnia and Herzegovina has established a state wide power transmission company (Transco) and an independent system operator. Regulation of the electricity sector is in place with the introduction of regulation at the state level for Transco and ISO and at the entity level for the generation and distribution and supply activities of the EPs. Progress is being made on the implementation of the action plans for further unbundling of the EP's generation, distribution and supply businesses. Accounts are being unbundled to be followed by legal unbundling and independent management of the sub businesses. All this is in compliance with the spirit of the EnC Treaty and the applicable directives.

Progress has been made in terms of market opening, though the actual implementation is still hampered by local tariffs that are lower than regional market tariffs.

In energy issues, according to the Constitution of Bosnia and Herzegovina, the entities should initiate all activities related to EE and RES while the role of the Ministry of Foreign Trade and Economic Relations coordinates this work and has an implementing role for all international programmes. The state cannot impose any policy or strategy on the entities, but can recommend, suggest and adjust the entities' policies and strategies for EE and RES.

Where problems exist they may often be traced back to the country's unusual constitutional arrangement. The state, while responsible for policy formulation coordination and interface with the international and regional bodies, is weak in terms of implementation capability. All assets are owned at the entity level and all implementation of policy is conducted at the entity level. And, at times, there appears to be ambiguity regarding the powers of the different levels of government.

Existing regulations and policy in the engineering, construction and building industry do not encourage energy saving (such as greater use of construction insulation materials and more cost-efficient heating systems). The majority of the (reconstructed) buildings dissipate a lot of energy and are less efficient than required by EC standards.

In November 2008, Bosnia and Herzegovina's leaders reached an agreement on energy policy principles.

There are no specialized agencies for energy efficiency or renewable energy. It is expected that the renewable sources sector will be arranged through a set of by-laws.

National priority areas for EE and RES

National priorities for the development of EE and RES will be developed during the formulation of energy policy and strategies. This work is now under way and is expected to be strongly influenced by the international obligations of Bosnia and Herzegovina as well as energy security considerations.

Interest in receiving equity and mezzanine financing

In general, interest has been expressed by the public and private sectors in the Equity Fund. The Ministry of Foreign Trade and Economic Relations (MOFTER) has indicated that the state is interested in the project and three main aspects should be taken into consideration for project implementation:

- Project activities should be coordinated at state level through the National Coordinator (MOFTER). This is important in order to ensure the success of the project.
- Energy efficiency and renewable energy projects development and implementation come under the entities only. Legal framework development is still in its first stage.

- Energy efficiency and renewable energy projects need to be implemented in reality in order to lead the entities to getting involved in the Fund.

MOFTER demonstrated a good interest for the Fund and summarized the situation as a good opportunity for the EE and RE sectors to benefit from the Equity Investment Fund and to promote the efficiency at entity level. The Fund could bring a good leverage to removing the financial barriers related to project implementation.

The Ministry of Finance is interested in all projects related to environment, but the non-existence of incentives for implementing EE and RES projects demonstrates the lack of resources and difficulties in having appropriate funds.

In the Republica Srpska, the Ministry of Economy, Energy and Development (MEED), showed great interest in the project. This Ministry is headed by a former minister of MOFTER who has a good understanding of the project and of its benefits. He stated that the Ministry was very interested in the Equity Investment Fund and was willing to provide full support for the project continuity. The conclusion that might be drawn from the meeting with the MEED is that the project is seen as a good opportunity and the entity is willing to put the necessary effort in order to benefit from the Equity Investment Fund.

The assessment mission met many representatives from the Republica Srpska in addition to MEED: the Regulatory Commission, the Faculty of Mechanical Engineering, the Chamber of Commerce, the Municipal sector representative, including district heating. All welcomed the Equity Investment Fund and indicated their willingness to collaborate with the necessary effort to seize all opportunities that it would bring to the region.

Barriers to financing EE and RES

Since Bosnia and Herzegovina is at an early stage of developing energy legislation with specific attention to EE and RES, special attention needs to be dedicated to overcoming the main barriers to promoting energy efficiency and creating a renewable energy market. Listed below are some of the issues that need to be addressed in the development of new legislation as well as in future work of the governmental organizations dealing with energy:
- Lack of awareness on EE and RES project benefits and advantages for the country;
- Lack of definitive studies about the potential of EE and RE at the country level;
- Lack of specific policies and/or strategies related to EE and RES at the state and entities levels;
- Weal coordination between the ministries, responsible for energy issues at the state and entities levels;
- Weak coordination and cooperation between the entities;
- Creation of a dedicated unit or department dealing solely with EE and RES within the ministries should be considered;
- Energy Agency at a state and/or entities level should be created;
- Lack of resources in the ministries, state and entities levels;
- Low energy prices and potential for feed-in tariffs;
- Necessity for harmonization of energy prices and regulation between neighbouring countries;
- Lack of defined methodology for concession and permit delivery for small hydro projects;
- Need for improved administrative procedures for small hydro project development.

Numerous requests for construction of small hydropower plants have been made but progress in actual construction is limited, mainly due to lack of a relevant framework for their actual implementation and also limited capital availability.

The price of electricity generated in small hydropower plants is considered to be very low by developers, and it prevents investors from further plant construction activities. A decision on methodology to determine a purchase price level of electric power from RE with installed power up to 5 MW has been adopted (Official Gazette FB&H 32/2002, Official Gazette RS 71/2003).

Power utility companies in Bosnia and Herzegovina are required to take over the electricity produced from RES. Current prices for renewables are:

- Small hydro plants: 4.45 euro cents/kWh
- Plants using biomass: 4.28 euro cents/kWh
- Wind plants: 5.56 euro cents/kWh
- Solar plants: 6.12 euro cents/kWh

There are several ongoing project activities related to RES, but more on an ad hoc basis and in a sporadic way. There is no systematic approach and institutionalization of procedures for RES related projects development to avoid their development on a voluntary case-by-case basis. Issues related to the network connection of RES also need to be resolved.

Incentives

No major incentives are available in Bosnia and Herzegovina, other than the feed-in tariff discussed above. However, incentives for tax reduction and/or exemption for equipment related to the implementation of projects could be foreseen since it has been already granted to the electric utilities and could be extended to other EE/RES projects.

Financing schemes

In Bosnia and Herzegovina the financing environment for EE and RES is still at its preliminary stage and no existing operational funds are accessible at the state or entities levels. The budgets available for addressing EE issues are very limited and compete with other critical areas of the economy for limited funds. However, the EE/RES sector is receiving considerable attention at present from the international community. These initiatives are addressed in below.

The Government of the Federation of Bosnia and Herzegovina has established a Fund for Environmental Protection[4], the purpose of which is to finance and support projects reducing emissions and improving the environmental conditions. EE and RES projects could be financed through this fund. It is foreseen that the fund will be endowed from penalties for pollutants (vehicles, industry, energy plants, etc.). A similar situation in terms of financing EE projects exists in the Republica Srpska. The Fund is not yet fully operational.

The main financing sources for EE and RES, for the time being, are regular commercial loans that investors do not consider favourable since the interest rates for long-term loans reach up to 12 per cent. The difficult financial situation of many companies in the industry and services sectors, combined with the low level awareness, do not make EE investments a priority.

It is likely in the near term that EE/RES projects will be funded through a mixture of grants and loans. Grants are available for project preparation while loans are needed for project implementation. Examples are the IPA 2007 project (Instrument for Pre-Accession Assistance supported by the European Commission) and EBRD initiatives discussed below, as well as the UNECE project.

---

[4] Regular review of energy efficiency policies 2008, Energy Charter protocol on energy efficiency and related environmental aspects.

ESCOs

There is no ESCO in Bosnia and Herzegovina.

Banking sector

From the banking sector, the Raiffeisen Bank is very interested in the energy and industrial sectors and shows a clear interest in the Fund. It is the only bank in Bosnia and Herzegovina known to be engaged in EE and RE projects in the country. This private bank is present in all the neighbouring countries and has initiated, with EBRD, some actions related to EE in the residential sector.

Other banks, especially with European headquarters, have expressed interest in the EE/RES area but meetings with them were not organized in the framework of the assessment mission.

Brief conclusions and recommendations

- Private and public sector investor interest is generally encouraging
- The preliminary stage of the legal framework, regulations and a lack of encouraging conditions and incentives are among the main barriers to financing EE and RE
- Any financial opportunity will help the development of the EE and RE market in Bosnia and Herzegovina
- For the private sector, the interest in RES projects is particularly high, especially regarding small hydro projects
- Most EE projects are likely to be in the public sector at the entity and municipality levels (district heat, street lighting, insulation, metering etc.).

Energy efficiency and renewable energy sources project development and finance capacities

Existing and prospective EE and RES projects

In Bosnia and Herzegovina, during the last years more than 100 concessions for small hydropower plants projects have been issued. However, only a few have been implemented so far. About 50 per cent could not be implemented because of the lack of investment resources.

Apart from the UNECE project, a number of EE/RES projects are in the process of getting underway in the country. These are listed below.

EC IPA 2007 and KfW: Terms of Reference have been prepared and are expected to be out for tender during early 2009 with award later in 2009. The project has four main components and provides technical assistance:
- Demonstration projects for energy efficiency/renewable energy
- Institutional and technical capacity building (single component)
- Public education
- Legal framework

Ten small demonstration projects are included in the project and feasibility studies for two larger projects (about €5 million each) will b e prepared. Funding for executing the larger projects is planned to be provided by KfW. The technical assistance (TA) budget is over €2 million.

EBRD: Western Balkans Sustainable Energy Direct Financing Facility: € 50 million fund, plus incentive grants, is available for promoting RES and EE projects for the demonstration of new replicable behaviour and activities, demonstration of new financing mechanisms, transfer of skills, legislative reforms and capacity building.[5] Funds are directed at RES projects and EE in the industrial rather than public (municipal) sectors.

The World Bank has in operation a regional project to provide direct loans for EE upgrades of infrastructure, market development support through EE fund structures (credits and guarantees), energy service company (ESCO) development and carbon finance support, for example through Green Investment Schemes. This project is established in a number of South-Eastern and Eastern European countries but, to date, not in Bosnia and Herzegovina.

Norway (Norsk Energi): The proposal is for developing greenhouse gas emissions reduction projects in the heating sector in Bosnia and Herzegovina, with the objective of assisting and training local municipalities and district heating companies to prepare projects for funding using the CDM. The project would have three phases; (i) inception, (ii) training local experts in project preparation and CDM requirements, and (iii) providing technical assistance for selected projects.

USAID SynEnergy Project: Capacity building and institutional network development is a regional project in scope and aims at advancing and strengthening the energy efficiency and renewable energy sectors of Energy Community (EnC) Treaty countries, promoting the rational use of energy and developing technical and institutional networks to support planning, policymaking, programme development, and implementation..

The beneficiaries of this project are mainly national energy agencies or institutes of the involved countries which contribute significantly to the implementation of national energy policies.

The activity will be developed to enhance the capacity of the energy agencies/institutes and establishing an effective network of policy- and decision-makers, public administrators, the private sector and civil society.

Assessment of investment project development skills

During the last decades Bosnia and Herzegovina has focused mainly on reconstruction and rehabilitation of the power sector, concentrating on the power plants, transport and distribution systems. EE and RES were not priorities of the work plan at that time and this explains the actual situation in terms of knowledge, legal framework, regulations, and skills for the identification and development of EE ad RES projects. Therefore, the mission assessment for the investment project development skills targeting EE and RES projects shows clearly that there is a strong need for support and capacity building in this area. This assessment is generally supported at the government levels.

Public sector

At the state level the Ministry of Foreign Trade and Economic Relations (MOFTER) is deals with the limited resources which would specifically deal with EE and RES matters and are reflected by a limited advancement related to EE and RES issues. Only a few officials deal with all energy activities within the Ministry. The Ministry does not have specialists to conduct and evaluate preliminary energy audits or investment grade audits. MOFTER stated

---

[5] 2008 projects: Western Balkans Sustainable Energy Credit Line Facility and Western Balkans Sustainable Energy Direct Financing Facility at www.ebrd.com.

that there was a need for skilled resources for EE and RES financial engineering and business planning.

At the entities level (Ministry of Economy, Energy and Development (MEED) of the Republica Srpska) it appears from the meetings held in the Republica Srpska that the situation is similar to the state. There is no allocated unit or group for EE and RES activities within the Ministry, though this is in the process of being formed. MEED recognizes that tasks go beyond the capacity now available.

There are few existing internal resources familiar with project development procedures, energy conservation measures, financial engineering and bankable project presentation and development. The absence of integration of EE and RES issues in the global planning process is the consequence of complex procedures, lack of knowledge, resources availability and some lack of understanding of the benefits and impact of developing EE and RES related plans and programmes.

MEED expressed its willingness to benefit from other countries' experience to improve the situation and endorsed the need for capacity building at all levels. The lack of EE agencies to implement and monitor EE and RE initiatives in the state and entities levels represents a major barrier toward improving the issues related to energy conservation and sustainable development.

According to the information gathered from stakeholders, it seems that the situation in the Federal Ministry of Energy, Mining and Industry in the Federation of Bosnia and Herzegovina is similar to the Republika Srpska. It has to be noted that no meetings could be arranged at the Ministry.

In the municipal sector, despite the USAID municipal energy management project that lasted four years in Bosnia and Herzegovina, the general point of view of most participants is that additional capacity building for the municipal staff in the administrative and operation and maintenance levels are still needed. Big municipalities could have the necessary skills and resources to identify energy conservation measures but not enough skills for project development and bankable project presentation. Municipal staff could have technical skills for identifying energy saving opportunities but not enough knowledge of methodologies to develop and make necessary assessment of the saving potential and financial needs (cost/benefit analysis).

For RES projects, the needs for technical and financial proposal development including district heating projects are even more significant.

On another level, permit issuing problems and the absence of methodology and guidance for RES approval and development constitute basic barriers related to developing such projects at the municipality and local community levels.

Private sector

Skills for energy auditing are still lacking in the Bosnia and Herzegovina market. The Centre for Economy, Technological and Environmental Development confirmed that the EE market is not sufficiently established to drive EE/RES activities and project development. There is a great need for capacity building related to project identification, evaluation and financial analysis. In many cases, the private sector does not have the capacity to identify EE and RES projects and to prepare bankable project proposals. In Bosnia and Herzegovina, there is no private company with a solid record for energy audit development or investment grade audit experience in the local market.

The main problem resides in the lack of methodology, understanding and specific requirements for investment grade audit development. Surveying procedures, energy balance development, risk assessment, measurement and verification procedures, financial engineering and energy management will be a part of the skills needed for bankable project development and presentation. The absence of these skills, along with tariff issues, partly explains the absence of ESCO in the market.

The commercial financial institutions do not have the required skills for EE and RES projects evaluation. It has been mentioned during the meeting with USAID, EBRD, World Bank and UNDP that the lack of EE and RE project understanding by the financial side reflects the lack of available financing with appropriate incentives. The Raiffeisen Bank constitutes an exception in the market and is already involved in EE and RE projects (USAID and EBRD). It considers EE and RES projects as any regular investment project, unless when offering a preferential rate on a few occasions.

Assessment of equity and mezzanine financing business development skills

Only financial institutions have knowledge about equity and mezzanine fund specificities. Other stakeholders do not have a specific understanding of the mezzanine and equity fund concept. A limited number of companies are able to play the role of project developers since there is no registered experience with equity and mezzanine finance in the country. No records have been found for third party financing, ESCOs, or project bundling for EE/RES project promoters. The potential project developers in the field of EE/RES have limited experience with the preparation of equity or mezzanine finance participation. The larger companies, such as electricity utilities, gas, petroleum, and large district heating companies, are more likely to have the necessary resources for project development based on the equity and mezzanine finance mechanism.

Capacity building on mezzanine financing mechanisms would be required, if such financing is to be tapped as a potential source.

Brief conclusions and recommendations

The assessment mission identified that technical engineering skills related to energy projects are available in the country. At the same time, following the assessment and according to the described situation in the above sections, the capacity building training for EE and RES project promotion and market creation would be very valuable in Bosnia and Herzegovina. In addition, the results of the capacity building programme that will be provided within the framework of the project would be enhanced if additional training could be provided on energy auditing and on the identification of energy conservation measures.

Investor interest

Public sector

The State Ministry of Finance would consider possible participation in the Fund when more details on its functioning are provided. It should be noted that the state may not have enough budget to cover the resources required to handle EE and RE issues, and probably less to participate in a fund that will not exclusively be built for Bosnia and Herzegovina. The conditional participation linked to the Fund's location, conditions and operation procedures will drive the government interest in participation in the Fund. The governments (state and entities) are short of financial resources and therefore do not have significant capacity to invest in the Equity Investment Fund.

The Ministries of Finance of the Entities may also be considered as potential investors as they run the Entity budgets. At present, no interest has been expressed by the

Republica Srpska for investing in the Fund. Lack of resources and budget were cited as obstacles.

Private sector

The private sector shows little interest at present, partly due to the lack of legal frameworks and regulation, but also due to limited local exposure to the potential of the EE/RES opportunities in Bosnia and Herzegovina. Also, existing financial incentives such as feed-in tariffs and tax incentives do not encourage participation in developing EE/RES projects. Raising awareness in this area would likely result in increased interest.

The Raiffeisen Bank showed an interest in participating in the Fund. However, discussions will most probably need to be held with its head office in Vienna.

Potential partners for project co-financing

The Raiffeisen Bank might be a good co-financing partner for the Fund. The partnership could be on two levels: cooperation in promoting the Fund in the region and provision of debt to the project implementation. Other commercial banks might follow, once real projects get under way.

Brief conclusions and recommendations

The interest of public and private sector investors observed during the mission for the Equity Investment Fund is low. The main reasons reside in the lack of legal framework and supportive programmes and incentives for the EE and RE sectors and limited awareness with respect to EE/RES project opportunities

Some barriers may be removed by developing a legal framework and regulations for EE and RES. Standardized procedures and requirements for developing EE and RES projects would also help.

BULGARIA

Energy Overview

Existing energy resources, energy dependence for primary and secondary energy resources, production of electricity and of heat, use of renewable energy sources (RES)

The period following independence in 1991 has been characterized by a significant decrease in energy use in the country. This reduction in energy consumption is not a sign of improved energy efficiency, but is due to many other factors inherent to the transitional period. Final energy consumption in 2005 was still 40 per cent lower than in 1990.

Despite this large decrease, Bulgaria's economy still has a high level of primary energy consumption per unit of GDP compared to averages for countries in the OECD. According to the International Energy Agency (IEA), in 2005, the energy intensity of Bulgaria (0,32 toe/$100O at ppp) is 2 times higher than the EU-27 average (0,16 toe/$100O). This high energy intensity is explained by the extensive use of electricity in metal processing industry; the low efficiency of electricity generation, supply and consumption; and the extended use of electricity for heating by residential and tertiary users.

Electricity generation is largely based on coal and nuclear energies (respectively 52 per cent and 42 per cent of the total electricity production in 2005, the remaining 6 per cent being hydroelectricity).

Currently there are 16 systems for central district heating in the country and district heating is the principal form of heat supply in the multifamily buildings in the bigger cities. The installations are 20-36 years old and the principal fuels are coal and gas. Electricity use for heating is decreasing; however, increased penetration of small air-conditioners is evident in cities. Natural gas has a very low penetration to date. The largest increase occurred for combustible renewable and waste and this is most likely due to the increased use of firewood in the residential sector.

The industry is the greatest energy consumer (32 per cent of total final consumption), followed by transports (27 per cent) and residential (20 per cent) sectors. While energy consumption is decreasing in industry, it is increasing rapidly in transports.

Consumption of final energy by industry has reduced by 60 per cent since 1990, influenced by the re-structuring and modernization of the Bulgarian economy. However, industry still has the highest share of final energy, mainly due to three energy intensive sub-sectors (chemicals and petrochemicals, non-metallic minerals, and iron/steel), though they have been largely modernized. The use of district heating has decreased sharply since 1997 leaving a place for electricity, natural gas and oil products. Coal resisted due to the continued presence of the steel industry based mainly on oxygen steel in the country.

Transport is the only sector where energy intensity has increased in the last ten years. There has been a strong modal shift from passenger rail to road transport and similarly from goods transport by water and rail to transport on roads. A similar shift occurred from public urban transport to private cars implying a constant growth of urban traffic congestion. The fuel mix has shifted towards diesel following the general patterns in Europe. There was also a strong increase in LPG demand.

Residential energy use decreased from 1990 to 2000, but it has started to grow again since. The share of each energy carriers has changed considerably: electricity is still by far the largest energy source, followed by district heating, but their shares have decreased gradually due to an increased use of wood for heating purposes. About 58 per cent of the dwellings are heated with individual room heating (including electric heating) and around 42 per cent have central heating (including district heating). The share of electricity used for heating purposes remained roughly at 25 per cent over the past 15 years.

Energy consumption in the services sector decreased from 1990 to 1997 but has increased again since. Electricity has by far the largest share in this sector, followed by district heat. Other sources such as oil products or natural gas make much smaller contributions.

Level of priority given to EE and to RES in the country's energy policy

During the last years the process of harmonization of the energy efficiency framework of Bulgaria with European legislation has been a priority. Energy efficiency activity is a matter of high priority at national level and increased attention is being given by national authorities to energy efficiency issues.

The Energy Efficiency Act was adopted by the Parliament in February 2004. The secondary legislation for the implementation of the Energy Efficiency Act is being developed. The latest ordinances that have been elaborated are mainly directed to the implementation of the legal provisions concerning the certification of buildings. A new energy efficiency law is under development.

Regional and local authorities started being legally involved in energy efficiency activities with the promulgation of the Energy and Energy Efficiency Act in 2002. This, together with a number of internationally-funded programmes and projects, gave an impetus to several local energy efficiency initiatives.

Investments in the energy sector

Bulgaria has a competitively diverse energy mix with an average level of dependence on imported fuels (oil and natural gas from the Russian Federation and also solid fuels). Domestic energy production includes nuclear energy and solid fuels which comprise the main fuels for electricity generation.

Although Bulgaria has substantial reserves of lignite, these reserves are difficult to reach (located under cities and villages), and are of low quality (high sulphur content). Natural gas reserves are limited and almost all the gas consumed in the country is imported from Russia. Bulgaria has modest hydroelectric resources. The wind and solar energy resource potential is significant, but only a few projects exist at the moment. Bulgaria has a sizable reserve of geothermal energy and is rich in low enthalpy geothermal waters.

As part of the agreement on EU membership, four (out of six) units of the Kozlodoui nuclear power facility were closed in 2003 and 2006. Since the beginning of 2007, units 5 and 6 have remained in operation, accounting for nearly a third of the electricity production. There are plans to restore nuclear capacity through construction of a new plant.

Bulgaria is working on promotion policies for ESCOs and policies to develop private funds. The main funding sources for energy efficiency services continue to be public sources but the frame of the EU Directive for Energy Efficiency and Energy Services has opened up chances for private energy services to develop further.

Financial environment in energy efficiency and renewable energy

Legal, regulatory and policy framework

The main legislation in the areas of energy and EE are the Energy Law, Energy Efficiency Law and a Draft of a new Energy Efficiency Law, secondary legislation to the above-mentioned laws.

The major policy documents (strategies, programmes, action plans) in the areas of energy and EE include the following:
- Energy strategy
- National long-term programme for energy efficiency 2005-2015
- First national action plan for energy efficiency 2008-2010.

The main legislation and major policy documents (strategies, programmes, action plans) in the area of renewable energy sources (RES) include:
- Renewable and alternative energy sources and biofuels law
- Secondary legislation to above-mentioned law
- Long-term programme for promotion of biofuels consumption in transport sector 2008-2020
- Long-term programme for utilization of biomass in Bulgaria 2007-2015
- National long-term programme for promotion of RES utilization, 2005-2015.

Given the fact that many of the measures have just been introduced in Bulgaria, relatively little is yet known in terms of impacts pointing to the necessity for Bulgaria to closely monitor their impacts in the coming years. The monitoring requirements laid down in the EU Directive 2006/32/EC for energy efficiency and energy services should help to develop further the required monitoring tools.

Regarding renewable energy, Bulgaria has good opportunities to exploit indigenous renewable energies although the current penetration of renewables is still very low. The target to be achieved in 2010 is about 11 per cent for electricity consumption and 16 per cent in 2020 (see draft EU Directive). The main potential for renewable energy production is from biomass, wind, geothermal and solar sources.

Interest in receiving equity and mezzanine financing

In Bulgaria, capital market development faces particular challenges arising from banking sector dominance and finding local comparative advantages in an environment of increasing capital market integration in Europe. At this point, though growing in importance, the Bulgarian capital market and non-bank financial intermediaries remain underdeveloped compared to the banking sector, similar to most Central and East European countries. Nevertheless, recent developments (including a boost by the Bulgarian National Bank's credit limits imposed on the banking sector in 2005/2006) and demand and supply incentives indicate that there is potential for the development of the private debt and equity markets in the medium term. The sustainable development of these markets, however, requires addressing some key bottlenecks, especially in the investor base and the stock market.

The Bulgarian Stock Exchange has intensified efforts to attract promising Bulgarian companies to the stock market through outreach and education activities, improvements in financial disclosure and corporate governance guidelines, and upgrades in trade transparency, infrastructure and supervision. There is significant potential for the listing of medium-sized companies and raising new capital by listed companies as already manifested by increasing the number of initial public offerings (IPOs).

The current stage of private debt and equity market development, as well as institutional investor size consistently put Bulgaria in the middle range among Central and East European countries. Bulgaria's ranking in private debt market size is ahead of several more advanced countries in the region. As in most Central and East European countries, corporate bonds have been predominantly issued by the financial sector. The nonfinancial corporate sector has not relied extensively on the bond market for funding given the strong competition in the banking sector. The largely foreign-owned and well-capitalized banking sectors of Bulgaria have been in a good position to provide the necessary funding to the corporate sector at competitive interest rates. For non-bank financial institutions, bond financing is likely to remain an important funding source, while non-financial corporates continue to have weak incentives to issue domestic bonds.

As a result, the assessment mission showed that the concept of equity and mezzanine financing is not widely known by EE and RES actors.

Barriers to financing EE and RES

Bulgaria experienced a decade-long delay in its transition to a market economy. In early 1997, the country experienced a severe economic and financial crisis, involving a sharp decline in GDP and per capita incomes, the collapse of the banking sector, and a major foreign exchange crisis. Following this crisis, the Government adopted a comprehensive economic reform programme supported by international financial institutions and other development partners, including major trade and price liberalization, social-sector reform, and restructuring of the financial, enterprise, agriculture, and energy sectors, including the divestiture of state-owned enterprises.

For the moment, energy efficiency is still not a matter of great concern in industry. By law, producers of goods and services with annual energy consumption equal to, or higher than, 3 000 MWh are subject to mandatory audits every three years. However, according to the Energy Efficiency Agency, many companies do not respect this obligation as there is no penalty defined by law. Furthermore, there is no constraint for energy consumption improvement and few audits result in actions. The problem comes especially from SMEs, with low technology level, low awareness about advanced technologies and little interest in energy efficiency investment opportunities.

Over 92 per cent of the residential building stock in Bulgaria is privately owned and most are owner-occupied. Nearly 40 per cent of the dwellings are situated in large-panel apartment blocks. Still, no widely recognized legal basis for associations of home owners exists, and this has led to low levels of management of the buildings and has hindered the attainment of a 100 per cent agreement among the owners on the implementation of energy efficiency measures in an entire building.

A significant barrier to the implementation of energy efficiency programmes in municipalities continues to be the lack of specialized sources for their financing in Bulgaria. Another significant barrier to energy efficiency projects is the weak interest of foreign investors in such projects. This was the result of the specific characteristics of this kind of investment projects – relatively small volume of the investments, lack of clarity with respect to the ownership on the sites, distorted baseline levels for comparing achieved savings and an absence of state guarantees.

Incentives

Bulgaria has put in place a comprehensive system of incentives based on feed-in tariffs which yields good results. The preferential tariffs set by Bulgaria are comparable with the German values, inferring that the tariffs are likely to be sufficient to promote renewables in Bulgaria. Financial support programmes for renewable energy have also been put in place. Recent developments show that the market is responding to the promotional regime.

The Energy Efficiency Agency proposes grants to SMEs covering 50 per cent of the costs of energy efficiency audits. This programme was financed with 5 million leva/year during three years, but only a part of this budget was allocated, due to a lack of participants. For the moment, energy efficiency is still not a matter of great concern in industry but a new energy efficiency act should soon define penalties for non respect of the mandatory audit.

The main general programmes for energy efficiency in residential and tertiary sector buildings are the National Programme for renovation of multi-family buildings (2006- 2020) and the National Strategy for financing of buildings insulation for energy efficiency improvement (2006-2020). In addition, Bulgaria has introduced a number of important energy efficiency measures in buildings, among them minimum efficiency standards, mandatory energy labeling and building performance standards. Also the Bulgarian Government has introduced individual billing in dwellings.

Financing schemes

EBRD manages two credit lines dedicated to energy efficiency in Bulgaria: the Bulgarian Energy Efficiency and Renewable Energy Credit Line (BEERECL) and the Bulgarian Residential Energy Efficiency Credit Line (REECL).

A similar fund should be established but EBRD does not have a specific project at the moment.

BEERECL was set up in 2004 by EBRD together with the Kozloduy International Decommissioning Support Fund (KIDSF) and the Ministry of Energy. The framework comprised €50 million of EBRD financing onlent to s ix participating banks, and was extended in 2006 with EBRD making available a further €55 mi llion. The financing is complemented by €20 million in grant funding from the KIDSF, set up in 2000 and administered by EBRD. Under this framework, technical assistance is included free of charge for the beneficiary of the project (conducted by EnCon services who won the EBRD tender).

To date, the Bulgarian Residential Energy Efficiency Credit Line (REECL) Programme has committed to 15,560 energy efficiency home improvement projects, financed through personal loans totalling 45 million leva and incentive grants amounting to 7.8 million leva, and saved a total estimated electricity equivalent of 108 GWh per year.

The Bulgarian Energy Efficiency Fund (BgEEF) was established in 2004, initially capitalized through grant financing, the main donors being GEF, the Government of Austria, the Bulgarian Government and private Bulgarian enterprises. The manager of the Fund is a consortium of three organizations: the Canadian energy efficiency consultancy "Econoler International", the Center for Energy Efficiency EnEffect and the non-banking financial institution "Elana Holding" plc. The underlying principle of the Fund's operations is a public-private partnership. The Fund pursues an agenda fully supported by the Government of Bulgaria, but it is structured as an independent legal entity, separate from any governmental, municipal or private agency or institution.

BgEEF supports the identification, development, and financing of viable energy efficiency projects implemented by Bulgarian private enterprises, municipalities and households. The Fund has the combined capacity of a lending institution, a credit guarantee facility and a consulting company. It provides technical assistance to Bulgarian enterprises, municipalities and private individuals in developing energy efficiency investment projects and then assists their financing, co-financing or plays the role of guarantor in front of other financing institutions.

The World Bank is rather satisfied with the design of the Fund. Although the BgEEF does not benefit from grants from KIDSF for energy audits, as the EBRD credit line does, it is still very attractive and currently demand exceeds the fund financing capacity. The capital base should be increased but the Bulgarian Government is reluctant to do so at the moment. The World Bank should be contacted for future ECE project activities, while assessing the Bulgarian institutional framework for the activities of an EE equity investment fund.

ESCOs

Enemona SA is a private Bulgarian engineering company, established in 1990. With more than 2000 engineers, it is the largest ESCO in Bulgaria. The company implements energy efficiency projects in industrial and power engineering buildings, based on energy performance contracting. Enemona also implements projects in municipalities' buildings but pointed out some of the difficulties they face. Some municipalities may prefer to rely on structural funds rather than on performance contracting, as no reimbursement is expected. Force of habit and the lack of trained personnel to identify energy savings potentials and conduct investment projects restrain the realization of viable energy efficiency projects. Finally, the delay passing the Ordinance envisaged under article 21, paragraph 7 of the Energy Efficiency Act, which is deemed to regulate the terms and procedures of determination and payment of fees under ESCO contracts in state owned and public buildings, also restrains the potential in public buildings. However, Enemona pointed out that Bulgaria has one of the more favourable legislations in Europe regarding energy efficiency.

Enemona established its first ESCO contract in 2004 (the first ESCO contract with industry in 2006). To date, Enemona has established 30 performance contracts based on the ESCO model. Enemona may be disposed to serve as a financing vehicle to address small energy efficiency projects (under €10 million) in industry and the public sector, while restricting transaction costs for an investment fund.

Banking sector

Like other booming countries (Estonia and Latvia), Bulgaria registers top scores in the EBRD ranking on corporate governance. Similarly, it was ranked higher on the business environment than other emerging European countries by the World Bank's Doing Business Survey.

However, according to the EBRD-World Bank Business Environment and Enterprise Performance Survey (BEEPS), in 2005, new investments in the corporate sector are mostly (65 per cent) financed on internal funds. The median debt to asset ratio is still low, suggesting that firms are not becoming more leveraged.

Bulgaria's commercial banks dominate the financial sector. Bank assets are more than ten times larger than total assets of the next largest financial subsector. The privatization of the banking sector is complete with most assets in the hands of foreign-owned institutions. Financial sector vulnerabilities are limited, given that the level of capitalization of banks is high and portfolio quality is good with only 2.2 per cent of loans overdue by more than 90 days.

The effects of the crisis of 1996–97 were still visible in the early 2000s, as banks remained very conservative and risk-averse in lending. Regulation and supervisory oversight have much improved since the crisis. The authorities have taken a number of steps to eliminate obstacles to financial intermediation. The Central Credit Registry accessible to banks was made operational. Bank monitoring was enhanced by the adoption of consolidated supervision in July 2000. In January 2003 International Accounting Standards were introduced for banks and other financial institutions. Amendments to the insolvency section of the commercial code in mid-2003 also made it easier for banks to lend. Ensuring sustainable credit growth is now a challenge. Due to recent strong domestic credit growth and increasing competition, domestic credit to the private sector now accounts for 67 per cent of GDP. The authorities recognized the need to tighten the regulatory and supervisory framework for the non-bank financial sector, now well supervised by the Financial Sector Supervision Commission.

Brief conclusions and recommendations

- Energy efficiency activity is a matter of high priority at national level and increased attention is given by national authorities to energy efficiency issues.
- Given the fact that many of the measures have just been introduced in Bulgaria, relatively little is yet known in terms of impacts pointing to the necessity for Bulgaria to closely monitor their impacts in the coming years.
- However Bulgaria registers top scores in the EBRD ranking on corporate governance, new investments in the corporate sector are mostly (65 per cent) financed on internal funds.
- In Bulgaria, capital market development faces particular challenges arising from banking sector dominance and finding local comparative advantages.
- The assessment mission showed that the concept of equity and mezzanine financing is not widely known by EE and RES actors.

Energy efficiency and renewable energy sources project development and finance capacities

Existing and prospective EE and RES projects

The results of the BEERECL to date are excellent with default repayment, 39 projects financed, corresponding to €19.4 million in BEEREC L loans (cumulated projects size €33.3 million) and €1.88 million in incentive grants.

To date, the REECL Programme has committed to 15,560 energy efficiency home improvement projects, financed through personal loans totalling 45 million leva and incentive grants amounting to 7.8 million leva, and saved a total estimated electricity equivalent of 108 GWh per year.

Currently, BgEEF has the capacity to deal with 20 to 25 projects/year but with an average project size of €200,000 it is looking for new sources of financing.

One project was identified under the Joint Implementation. The project envisages the establishment of nine hydro powerplants on the river Iskar, about 40 km north of Sofia, with the overall objective to generate Emission Reduction Units, 370,969 tonnes of CO_2 equivalent in the period 2008 to 2012 (inclusive).

Public sector

The Ministry of Economy and Energy was established by decision of the Bulgarian Parliament in August 2005 through the merger of the Ministry of Economy and Ministry of Energy and Energy Resources. The Ministry pointed out two main issues that should be considered for future project activities. Experience from existing European financial mechanisms (Structural Funds, Clean Energy Fund, etc.) shows that, though there were large possible opportunities in Bulgaria and the technical expertise available at national level was good, the lack of awareness on existing financing mechanisms and the lack of expertise to apply were problematic.

Wind Power Station (Chernomorie – Enel)

Assessment of investment project development skills

Regional and local authorities started being legally involved in energy efficiency activities with the promulgation of the Energy and Energy Efficiency Act in 2002. This, together with a number of internationally-funded programmes and projects, gave an impetus to several local energy efficiency initiatives. Each year, Sofia municipality allocates funds for public buildings renovation, including energy efficiency measures but without necessarily high performance criteria. Sofia municipality may lack the capacity to deal with energy efficiency issues and most of this activity is subcontracted. Notwithstanding, a municipal energy agency (Sofena) was set in July 2001 under the SAVE II Programme of the European Union. Its main objectives include assistance to Sofia municipality in developing a sustainable energy policy, developing models for Sofia municipality energy planning and supporting their implementation.

Private sector

The Energy Efficiency Agency pointed out the large potential for EE in Bulgarian industry. Most industrial and tertiary actors are subject to energy audits but, until now, EE actions following audits are not mandatory (this may change with a new Bulgarian Energy Efficiency Act). The auditing and technical capacity exists, at least in auditing companies, but the lack of knowledge and understanding of energy efficiency issues by industrials restrains the realization of actions. There is also need for capacity reinforcement and training of experts on EE business plan development.

The Bulgarian Industrial Association is a voluntary commenced the large scope of activities for energy efficiency in industry but the existing energy efficiency fund is too modest to cover financing needs. The Bulgarian Industrial Association welcomed the ECE initiative. However, the typical project size suggested (€20 million) seems oversized for the Bulgarian market and most of the EE potential would not be addressed. Some intermediaries should be used to address smaller projects. There is a huge interest in equity financing in Bulgaria but mostly for renewable energy projects.

Even for the banks who manage EBRD's energy efficiency credit lines, evaluation procedures seem to focus on the clients' assets and collateral, as for a normal loan evaluation. The technical expertise on energy efficiency business plan evaluation is subcontracted and the bank lacks capacity on these issues. Currently, equity participation may not modify a client evaluation and capacity reinforcement may be useful.

The Center for Energy Efficiency EnEffect, founded in 1992, is a non-governmental, non-for-profit organization and its mission is to support the efforts of the central, regional and local authorities in the sustainable development of the country through more effective use of energy. In compliance with the Bulgarian Energy Efficiency Act, a consulting subdivision was established in 2004 – EnEffect Consult Ltd. Its main activities are in the field of energy audits, building certification, researching and analyses. Several EnEffect Consult activities are strongly in the scope of the UNECE project: energy and financial analyses, assistance in developing application documents for financing of energy efficiency projects, assistance in preparing documents related to GHG emissions trading, assistance in developing EU structural funds application documents. EnEffect already enjoy a serious influence on the energy policies of Bulgaria. EnEffect is the national coordinator for UNECE projects in Bulgaria and an incontrovertible actor for future project activities in the region.

The Technical University of Sofia is the largest higher engineering school in Bulgaria. In collaboration with the Norwegian company Energy Savings International SA and EnEffect, the Technical University provides training on energy efficiency auditing, adapted to the Bulgarian context. At the end of the training, auditors should be able to carry out complete energy audits and develop, in a proper and convincing way, reports to be used as the basis for the selection and implementation of energy efficiency measures, as well as preparing manuals and routines for efficient operation and maintenance, and for energy management of their specific buildings. The training includes business plan development. The Technical University also conducts energy audits with a team of 12 experts and supports other university experts. The Technical University could be consulted during future UNECE project activities, regarding regulatory framework analysis and training sessions.

Assessment of equity and mezzanine financing business development skills

The assessment mission showed that the concept of equity and mezzanine financing is not widely known by EE and RES actors. The specificities and interest of such mechanisms should be explained in detail in order to distinguish the fund from other financing mechanisms available. In addition, the assessment mission showed that technical and financial capacities are dispersed among actors: project developers lack financial capacity to present complete bankable project proposals and financial institutions subcontract entirely technical and business plan evaluations. These actors may benefit from a wider view and should gain at least a basic understanding of both technical and financial issues.

Nevertheless, the mission showed that some country experts in Bulgaria, namely BgEEF, Enemona, EnEffect and EnCon, may have the ability to serve as project developers, intermediaries that can work with the investment fund managers and have the business development skills to prepare equity or mezzanine finance participation in large companies, energy service companies (ESCOs), special purpose vehicles or other third party finance entities.

Brief conclusions and recommendations

The assessment mission showed that good technical skills exist in Bulgaria, both on energy efficiency and renewable energy issues. A number of different Bulgarian organizations have participated in EU funded projects, including regional and municipal energy and energy efficiency agencies, universities, NGOs, etc. Many of these projects have supported the capacity building of these organizations, and have helped to raise the awareness of the authorities and the general public. A number of local energy agencies were established with the support of the SAVE Programme. The Technical University of Sofia has already trained more than 450 energy auditors and several auditing companies are active in Bulgaria. Also, other programmes like the Municipal Energy Efficiency Programme aimed to enhance municipal and industrial capacity to develop bankable energy efficiency projects.

However, the assessment mission showed that there is a need for project development and financial project analysis capacity building. Beyond the existing capacity in Bulgaria to identify and assess potential EE or RES projects, the capacity to develop fully bankable project proposals corresponding to financial institutions standards and practices is rather restricted to a limited number of actors.

The funding sources for energy efficiency activities in Bulgaria have evolved considerably over the years, stimulated and supported by the EU accession process. While in the earlier years donor programmes played an important role, in recent years Bulgaria's own funding sources, for example from the state budget and, most importantly, the EU structural and cohesion funds, have taken up the major role. Bulgaria has also developed interesting funding schemes by way of public-private partnerships like the BgEEF and specific credit lines. The challenge for the future will be to enhance the reach and the size of these funding schemes.

Furthermore, a crucial issue would be to extend project realizations without such support scheme on a traditional business basis. The good results of ESCO companies in Bulgaria and the zero default repayment registered in both EBRD's credit lines and the World Bank's investment fund, as well as the good internal rate return (IRR) registered for such investments are encouraging. One of the main issues to be dealt with in order to attract private investors is the relatively small size of projects in Bulgaria. This issue could be addressed through dedicated financial vehicles, like ESCOs or intermediate investment funds which already exist in Bulgaria.

Investor interest

Public sector

The Ministry of Economy and Energy welcomes the UNECE initiative and is looking forward to future project activities. The Ministry is interested in the results of the tenders under the UNECE project and is looking forward to continuing discussions.

The Technical University of Sofia could be consulted during future UNECE project activities, regarding regulatory framework analysis and training sessions.

Private sector

Overgas strongly supports a diversification of energy sources in Bulgaria and especially promotes the switch from electricity to gas for heating. Overgas is interested in future project development.

CEZ [6] does not have a specific department dealing with energy efficiency and may need capacity reinforcement in the scope of EU directive 2006/32. CEZ may soon intend to implement projects in renewable energy production and would have to find co-financing.

The Bulgarian industrial sector commented on the large scope of activities for energy efficiency in industry and the existing energy efficiency fund is too modest to cover financing needs.

The Bulgarian Energy Efficiency Fund (BgEEF) also showed interest as it is currently looking for new sources of financing.

---

[6] Producer of thermo-power in the Czech Republic operating also in the Central and East Europe.

EnCon Services has experience in the development, finance and implementation of credit guarantee facilities for energy efficiency projects. EnCon has been involved in the development and implementation of energy efficiency and renewable energy projects in Bulgaria, and is also very interested in future project activities.

Enemona established its first ESCO contract in 2004 (first ESCO contract with industry in 2006). To date, Enemona has established 30 performance contracts based on the ESCO model. Enemona may be disposed to serve as a financing vehicle to address small energy efficiency projects (under €10 million) in industry and the public sector, while restricting transaction costs for an investment fund.

Potential partners for project co-financing

No potential partner for project co-financing could be identified at this stage.

Brief conclusions and recommendations

- The Ministry of Economy and Energy and the Ministry of Finance welcome the ECE initiative and are interested in future project developments (tender results).
- Public institutions and private companies already active in EE and RES are interested in taking part in future project activities (BgEEF, Enemona)
- Regarding the private sector and utilities, there is a large scope of activities for EE in industry and building stock but typical project size envisaged for the fund seems oversized for Bulgarian market.
- The interest is in equity financing in Bulgaria but mostly for RES projects.
- No potential partner for project co-financing could be identified at this stage.

CROATIA

Energy overview

In Croatia the share of renewable energy sources (RES) in gross electricity consumption reached about 34 per cent in 2006 (6,149 GWh, which is almost 50 per cent of domestic production). Renewable energy sources cover a large share of electricity generation in Croatia. This is achieved mainly by the significant number of hydroelectric power plants in the country. Total hydropower installed capacity was 2,056 MW_e in 2006, dominated by large HPPs, with hydropower generation accounting for 6,070 GWh in 2006. The Croatian wind sector has seen a large expansion in recent years, reaching 17.2 MW_e of total installed capacity in 2006, coupled with a generation level of 20 GWh. Other renewable energy sources have a low significance in the country's power generation; however, as a secondary legislation package has been approved for the support of renewable, the chances of increases in these areas are high. There is a relatively high wind potential for wind power in Croatia. Currently research is being carried out on the potential construction of wind farms of a total installed capacity of about 1,500 MW_e, however in order to maintain the secure operation of the electricity system it is very probable that only some of these projects will be realized. Croatia has significant potential for biomass and geothermal energy, with 50 PJ of total energy biomass potential and 48 MW_e potential capacity for electricity generation in the case of geothermal energy. Potential for additional hydropower utilization is not significant (possibly due to the already high utilization rate in the country), with a potential capacity of 177 MW_e for small hydropower plants.

Croatia is aware of the importance of energy efficiency as an integral part of demand management, and has begun to tackle most of the issues covered by the 2006 Directive on Energy End-Use Efficiency and Energy Services. Legislation covering most areas under the energy efficiency *acquis communautaires* has either been adopted or is under preparation. The Energy Efficiency Fund grants incentive loans for projects on the basis of public tenders. Croatia has not yet established a national savings target, an energy efficiency action plan or a regular system of energy audits of energy intensive industry.

Financial environment in energy efficiency and renewable energy

Legal, regulatory and policy framework

The European Energy Charter, signed by Croatia in 1993, implies the introduction of a long-term cooperation model in Europe in a market economy environment, which is based on cooperation between the signatories. The development of the energy market in Croatia formally began in July 2000 by initiating the Reform Programme of the Energy Sector in the Republic of Croatia. A year later the Parliament passed a set of acts on energy activities regulation which helped relations in the Croatian energy market, according to the EU directives. The Energy Law passed in 2001 regulates measures to ensure a secure and reliable energy supply, efficient power generation and its use. It addressed equally the enforcement of regulations in the energy sector, and regulates carrying out energy activities based on market principles or pursuant to public service obligation, and other key issues relevant for the energy sector. The principal objectives of Croatia's energy policy are stated in the Energy Sector Development Strategy adopted by the Parliament in 2002 for a period of 10 years. On the basis of the Strategy, a national energy programme (PROHES: Programme of Development and Organization of the Croatian Energy Sector) was developed. It was launched to develop an energy management framework that will promote clean technologies, shift to fuels with lower carbon content (natural gas), diversification of energy resources, higher EE and RES utilization, demand-side management, energy savings development of energy market, and environmental protection. Secondary legislation regarding electricity production from renewable energy sources was prepared and adopted in 2007. Secondary legislation regarding issues related to heat production and cooling from RES is currently under preparation.

The Government has started to reform the energy sector, in order to push the share of renewable energy sources (other than large hydropower plants) currently from one per cent to 5.8 per cent until 2010. The Energy Law has been supplemented by five regulations, which came into force in July 2007.

RES electricity feed-in tariffs are set according to the energy source from which it is generated. Green electricity producers that have signed a contract with the market regulator are eligible for these tariffs.

National priority areas for EE and RES

The national priority is the establishment of clearly defined national strategy and policy towards EE and RES, based on the introduction of a stable and harmonized legislative framework. Some secondary legislation for EE and heating/cooling from RES is still lacking and there are plans for it to be prepared and adopted in 2009.

Preparation of action plans for EE and RES is very important. The national strategy and legislative framework should define binding national targets, while action plans should present measures to achieve envisioned targets. The targets for EE and RES should be based on a realistic assessment of national potential, both in technical and economic terms, and on cost-benefit analysis.

Broader public awareness campaigns should be organized to increase private and public demand for EE and RES technologies and stimulate the market for these technologies. Furthermore, information centres, for technical and financial assistance, could be established at the national and municipal levels. Educational schemes for students and teachers, as well as training seminars for project developers, investors and decision-makers at the national and local level would be valuable.

Interest in receiving equity and mezzanine financing

The positive changes in the EE and RES market conditions are favourable for investment and the Investment Fund could be attractive for the implementation of some projects. At the governmental level such interest was expressed by representatives of the Ministry of Economy, Labour and Entrepreneurship. The Government is involved in various issues related to EE and RES projects development and implementation and is generally supportive of various instruments to make them possible.

Another governmental entity, the Environmental Protection and Energy Efficiency Fund (EPEEF) sees the Investment Fund as a complement to the different instruments already in place and its integration among other financing mechanisms will have a positive impact for project implementation.

Within the Chamber of Economy, the Energy Association has about 150 Croatian companies registered to perform energy related activities. The Renewable Energy Sources Affiliation is a sub-group within the Energy Association and has about 60 companies as members. The Association does not have resources to be involved in the Investment Fund financially but considers that opportunities for new projects will emerge for financing from it.

Representatives of the state-owned Croatian Bank for Reconstruction and Development (HBOR) believe that the Investment Fund might bring additional leverage for increasing the awareness level and better capacities for project implementation

HEP-ESCO, the governmental utility-based energy service company (ESCO) sees the Fund as an additional opportunity for the market to invest in potential projects that need this specific support. Up to now, HEP-ESCO, mainly oriented towards working with the government sector (schools, hospitals, other public buildings, municipalities, etc.), offers projects without a saving guarantee. The Investment Fund might bring a new opening to support the private sector projects where financing conditions and payback periods are tighter than for the government sector. HEP-ESCO interest might be relatively low for the Investment Fund targeted project size. The company has been implementing projects since 2004 but they are relatively small. So far HEP-ESCO has not been involved in energy performance contracting.

Representatives of EETEK, a private ESCO company, consider that the Investment Fund will bring a good opportunity to the private sector where financing mechanisms are limited especially for large-scale EE and RES projects.

The Energy Institute Hvroje Požar (EIHP) is a state-owned entity but is not funded from the government budget. EIHP is involved in a number of research and implementation activities in the energy sector. Its representatives reiterated the necessity of the Investment Fund for the Croatian market, and the need for more financial mechanisms to create better opportunities for potential projects in EE and RE.

Barriers to financing EE and RES

During the assessment mission many barriers were identified by stakeholders. The following are considered to be major barriers to EE and RES markets development:
- Lack of coordination between the ministries and government institutions
- Lack of financial and human resources at the national and local levels
- Lack of public awareness on energy and environmental issues, including EE and RES benefits
- Lack of public awareness about advantages of ESCO
- Lack of competition among ESCOs (only one state-owned ESCO established with support of the World Bank)
- Lack of legal framework for ESCO activities
- Lack of public-private partnership projects and initiatives
- Lack of experience in project financing, including EE and RES projects.

Incentives

The incentives for EE and RE projects are available in Croatia through the programmes and financing mechanisms promoting project development and implementation. There are also feed-in tariffs for electricity production from renewable energy sources.

EPEEF is an extrabudgetary governmental fund that provides additional resources for financing programmes and projects in the area of environmental protection, energy efficiency, and renewable energy sources. It can support up to 40 per cent of the total investment under the following conditions:
- Loan – zero interest rate, repayment period (payback five years, grace period two years), maximum EUR 230,000 per project
- Interest subsidy – 2 per cent decrease of the market interest rate (bank loan)
- Financial aid (grant) – only local/regional authorities are eligible.

Through the financing mechanisms for EE and RES projects, the Croatian Bank for Reconstruction and Development HBOR offers support of up to 50 per cent of the total eligible costs for the preparation of documents and studies for the renewable energy sources projects. In addition, for project financing related to environmental protection, EE and RES,

HBOR may offer a loan with interest rates in the range of 4 to 6 per cent depending on the borrower.

Another incentive offered by HBOR is an instrument to issue bank guarantees for EE projects only. The eligible projects are:

- New investments aimed at improving EE in buildings, i.e. in heating sources, local heating systems and heating networks where at least 50 per cent of the energy is used for maintenance of ambient temperature and water heating in buildings for household requirements.
- Greenfield projects, particularly those featuring integrated design and low energy building concepts, using high efficiency technologies/systems.

The EE project must have an estimated simple payback period of less than or equal to ten years. Bank guarantees are issued for the coverage of up to 50 per cent of total investments with a maximum of $300,000. Two commercial banks are included in the programme: Raiffeisen Bank Austria d.d., Zagreb, and OTP Banka Hrvatska d.d., Zadar.

Meeting at the Environmental Protection and Energy Efficiency Fund (EPEEF)

Financing schemes

EPEEF (described above) finances projects and activities in three basic areas: environmental protection, energy efficiency, and the use of renewable energy sources. In 2004-2007, it provided partial funding for over 270 projects related to EE and RES with a total commitment of €17.07 million.

HBOR offers two financing mechanisms for EE and RES projects (mentioned above).

The first is the Loan Programme for the preparation of renewable energy resources projects, which supports the preparation of documents and studies up to 50 per cent of the total eligible costs of design documents. The maximum loan is $ 150,000, with a grace period of up to one year and a disbursement period up to two years.

The second mechanism is the Loan Programme for the financing of projects of environmental protection, energy efficiency and renewable energy resources, intended for project implementation. There is no formal limit to the loan amount, but it depends on HBOR financing capabilities, investment project, creditworthiness of the borrower and quality of security offered. HBOR can finance up to 75 per cent of the estimated investment value with a maximum of one year for disbursement, a grace period of one year and repayment period of up to 12 years, including the grace period. The interest rate varies from 4 per cent to 6 per cent based on the borrower. Sixteen commercial banks are cooperating with HBOR on this loan programme.

However there is a lack of requests for these instruments from HBOR. HBOR representatives were not able to identify specific obstacles and bottlenecks that prevent wide use of these loan programmes in Croatia. Lack of specific EE and RES targets, skilled resources for EE and RES project development and a clear understanding of how these financial mechanisms work are the most likely causes.

HEP-ESCO, the utility based Energy Service Company, offers financing for energy efficiency at less than 5 per cent interest rate, with a maximum payback period of ten years for the governmental sector and five years for the private sector.

EBRD activities in Croatia are diversified and investment is mainly made in the infrastructure projects. The Bank sets an objective of promoting EE and focuses on infrastructure needed for security and diversity of supply, including renewable energy projects. In 2007, EBRD signed a commitment to finance up to EUR 25 million for the Private Equity Fund dedicated to investing in RES and EE projects in Central and South-Eastern Europe. The Fund invests in Central and Eastern Europe and South-Eastern Europe including Bulgaria, Serbia, Croatia, the former Yugoslav Republic of Macedonia and Ukraine, in projects qualifying as power projects that contribute to specific EU targets for the Kyoto Protocol, and have predictable and legally binding long-term off-take agreements. Regarding the structure, the Fund invests on its own or jointly in projects where equity returns can be enhanced by optimal degrees of debt using various financing structures.

The World Bank supported the establishment of HEP-ESCO in 2004. UNDP is implementing a GEF project "Removal of barriers for energy efficiency in Croatia". The main objective is to implement economically feasible EE technologies and measures in the residential and service sectors. The project is divided into two major activities: pilot project House-in-order and energy management in cities.

ESCOs

Two ESCOs are working in Croatia: HEP-ESCO and EETEK.

HEP-ESCO is owned by the state-owned utility company HEP and offers its services to all market sectors. However, almost all the projects implemented up to date are in the governmental sector. HEP-ESCO provides financing for the projects and does not offer a complete energy performance contracting scheme where payment is linked to the savings. The projects are justified by both energy bills reduction and equipment modernization and upgrade. HEP-ESCO's revenues have been growing steadily and were expected to reach about $ 10 million in 2008.

EETEK is a private company based in Hungary providing direct equity investments in Central and Eastern Europe. It is trying to offer ESCO services in the Croatian market. So far it has implemented one project in renewables (biomass) and one on energy efficiency (with a ceramics manufacturer, initial investment of $ 0.7 million). Lack of energy efficiency regulation and administrative barriers are cited as main obstacles to investment projects. In

Croatia, EETEK is looking for a project portfolio of up to $ 20 million per year where it can provide about 30 per cent of the investment.

Banking sector

Through the HBOR programmes a number of commercial banks are formally involved in the EE and RES projects as indicated above. However the level of utilization of existing financing mechanism remains very low.

Brief conclusions and recommendations

The financial environment in the country is characterized by existing opportunities and availability of financing for EE and RES projects. Investment resources for the private sector could be found through private equity funds and investment companies. However, utilization of existing resources, including financing mechanisms for EE and RES available in the governmental development bank, is low. Better information and awareness campaigns will certainly drive more studies support requests and demand for project implementation.

- Interest in the new Investment Fund could be assessed as moderate
- The Investment Fund could bring complementarities to the financing instruments in place for project implementation, including cooperation with the Croatian Bank for Reconstruction and Development (HBOR) and the Environmental Protection and Energy Efficiency Fund (EPEEF)
- The Energy Association of the Chamber of Economy could be a good partner for the Investment Fund as a potential source for finding viable EE and RES projects for implementation.
- HEP-ESCO may potentially be a partner for the Investment Fund as it has practical experience in financing and implementing EE projects in the country.

Energy efficiency and renewable energy sources project development and finance capacities

Existing and prospective EE and RES projects

The Ministry of Economy, Labour and Entrepreneurship indicated that the potential for EE and RE projects is significant. In 2007 about 244 requests for electricity generation from RES projects were registered with the Ministry.

The projects implemented or being considered by EBRD, the World Bank, UNDP and Croatian state-owned and private companies are described above.

Assessment of investment project development skills

Public sector

At the national level the lack of human resources and institutional capacity constitutes a problem. Skills for project investment development or detailed investigation of the technical and financial aspects are not required at the ministry level, however better understanding of these requirements would be helpful for developing of the appropriate legislation and policies. Project evaluation is part of the activities of another government body – the Environmental Protection and Energy Efficiency Fund (EPEEF).

The quality of the projects submitted for funding to EPEEF is uneven and this reflects the need for improvement of project development skills. The need for energy auditors has been estimated in the Master Plan for EE in Buildings at approximately 500 professionals. The Ministry of Environmental Protection, Physical Planning and Construction is planning to request the EPEEF to help in establishing a training centre for capacity building, including for training energy auditors.

The Energy Association of the Chamber of Economy, which deals with EE and RES project development and provides support, also need some capacity building. On the technical level, members of the association are well qualified. However, on the financial analysis level, risk management, monitoring and verification, project energy management they need additional training.

A template with a simplified procedure for bankable projects has been requested by a number of stakeholders.

At the municipal sector the regional energy agencies (developed under the framework for intelligent energy Europe programme) provide information related to EE and RES, as well as support and training related to energy management issues to municipalities. However, these newly established entities do not have all the required skills for investment project development. The regional energy agencies could be the main portal for cities to establish qualified teams and resources for EE and RES project development and disseminate information on financial mechanisms available at the national and international levels. Experts with good technical skills are generally available only in big municipalities. Even there, they do not deal with financial issues and complete project development. The same applies to district heating companies; the resources are more specialized in company systems, but do not have enough skills to develop new projects especially those related to EE and RES.

Private sector

In the private sector, certain skills for energy auditing exist but more capacity building with an emphasis on project measurement and verification of the savings/avoided energy and energy management is needed. Based on the HEP-ESCO experience, the market needs skilled resources in Investment Grade Auditing that masters all bankable project components. It is important to focus on the measurement and verification aspects for project development because most experienced people in this field are not familiar with the International Performance Measurement and Verification Protocol. The regular audits that are performed are not considered for investment purposes.

There is a lack of experienced project developers for EE and RES projects in Croatia. This is related to both lack of awareness on missing skills on the one hand and the market potential on the other. The lack of awareness about ESCO services and performance contracting prevents private companies from investing and developing bankable projects. With the existing ESCO companies in the country (particularly HEP-ESCO), the ESCO concept is largely misunderstood, as HEP-ESCO does not offer performance contracting services. Overall, the ESCO concept is not well known and this is another reason for the relatively low interest in EE and RES project development.

Investment project development related to EE and RES is not business-as-usual for banks and other financial institutions. More than this, promotion of the available funds specifically intended for such projects among bank clients is virtually non-existent. The case of existing financing mechanisms for EE and RES provided by HBOR that are almost not used, reflects the lack of awareness, promotion and cooperation between the stakeholders in the market. According to HBOR representatives, skilled resources for investment project development in general exist, however they are not at all involved in the EE and RES projects.

Assessment of equity and mezzanine financing business development skills

Equity and mezzanine financing is used by investment groups and private equity funds that are active in the region. However, the concept is not widely known by stakeholders in EE and RES project development. Equity and mezzanine financing should be explained to the potential beneficiaries of the project (both in the public and private sectors) in order to distinguish the Investment Fund from other available financing mechanisms. The integration of the proposed financing mechanism and its complementarity to the already existing ones would be valuable since it is important to put the Investment Fund in the country context. There is an overall impression that there is a missing link between the technical and financial capacities, which prevents full realization of the market potential. Project promoters, associations, regional energy centres and institutions would benefit from capacity building and training activities on business development skills for equity and mezzanine financing.

Large companies and groups such as district heating, hotel chains and large industrial companies are more familiar with the equity funds since they have been used during the privatization and modernization process. Insufficient knowledge and lack of practical utilization of the mezzanine financing mechanism suggest a need for capacity building on this issue.

Brief conclusions and recommendations

The meetings held during the assessment mission in Croatia and discussions with different stakeholders indicate that support is needed for capacity building, mainly oriented to higher level skills for project development and financing solution packaging. However, the need for raising awareness among the government institutions is recommended and the identification of the obstacles and bottlenecks preventing available programmes from realizing their full potential are also needed. This would make it possible to offer the market a complete solution package that would combine the advantages of both the new Investment Fund and already available financing mechanisms.

The project implementation rate is rather low considering all the investment mechanisms already in place. For RES projects this is explained by the transition period as project developers are waiting for the finalization and harmonization of all regulations to get full benefits. For EE, it has more to do with the lack of project promotion and insufficient awareness at all levels of the benefits of energy efficiency.

Encouraging the creation of new ESCOs is important since the number remains very modest in Croatia. In addition, existing ESCOs do not make available an Energy Performance Contract for potential customers. Support is required to assure good quality services and project success to enhance the actual level of ESCO concept comprehension and application. However, there is no indication at this point that market conditions are favourable for a new ESCO to be created. Particular attention is needed for relatively small projects, mainly related to EE. Capabilities for project bundling are needed to present bankable proposals to the Investment Fund, and ESCOs are generally considered as a major player in this field.

Investor interest

The possibility of the Croatian Government becoming an investor in the Investment Fund is low. EPEEF expressed its interest in investigating the opportunities to invest in the Investment Fund but taking into account its current mode of operations the probability of this is low.

HBOR expressed interest in getting more detailed information on the Investment Fund when it is available to consider the possibility of becoming an investor.

Private sector

No interest was expressed by the representatives of the private sector in investing in the Fund at this point.

Potential partners for project co-financing

HBOR and commercial banks participating in its loan and bank guarantee programmes could potentially be partners for projects co-financing.

There is a potential for cooperation with HEP-ESCO in project co-financing, particularly for private sector projects.

There is a possibility that specific EE and RES projects of interest to the Investment Fund could be co-financed by EPEEF.

Brief conclusions and recommendations

The concept of the Investment Fund is not well understood so far. More work is necessary to explain the modalities of its operation and potential advantages to stakeholders in the country. Potential for projects co-financing exists and should be further explored both with governmental and private entities. There is a low probability that there will be interest in Croatia in investing in the Investment Fund itself.

KAZAKHSTAN

Energy overview

Kazakhstan has large quantities of energy and mineral resources. The energy resources include coal (70 per cent), oil (22 per cent) and gas (8 per cent). Kazakhstan also possesses significant renewable energy resources including hydro, solar and wind energy. Approximately 50 per cent of the energy produced is exported.

Currently the power plants of Kazakhstan dispose of a potential capacity that can entirely meet domestic demand, but due to the existing structure of electricity transmission and the state of the market, the South and West Kazakhstan regions import electricity and capacity from the neighbouring states (2-3 per cent of the total electricity consumption).

Kazakhstan has a substantial electric power industry, the third largest in the former Soviet Union after the Russian Federation and Ukraine: 54 power generation plants with installed capacity of around 18 500 MW (88 per cent in thermal power plants and the rest in hydroelectric plants). More than two thirds of the electricity is produced by coal power plants. A substantial part of the coal is from open coal mines, and is very cheap. The coal is cheap even including costs for 2-3,000 km transportation from the coal mines to for instance heat and power plants in Almaty. There is a high potential of environmental improvements from purification of coal fired power plants; some of the coal used has 40-60 per cent ash content.

Transmission and distribution losses accumulate at about 15 per cent (technical losses).

Electricity demand has been steadily growing over the last years, and now all the power plants are operating at full capacity. This is why the Government recently has focused on the need of for energy saving, and initiated the process of preparing laws on energy savings and renewable energy. According to the Antimonopoly Committee, sufficient electricity will be produced up to 2015 (according to the approved State Energy Sector Development Programme this information varies to up to 2030).

Almost 100 per cent of the electricity production and distribution is privatized, while the national transmission company is state owned.

At present, the electricity tariff in Almaty is $ 0.07/kWh including VAT, for all consumers.

Since the electricity tariffs are low, very limited funds has been spent on maintenance and renovation of the electricity systems, and the investment needs are high. The total installed capacity is 18,500 MW, but due to old and inefficient equipment, only 14,000 MW can be produced.

The government policy regarding heat supply is directed to privatization. As stated in the Energy Sector Development Programme up to 2030, the development of centralized heating systems on the basis of co-generation plants where it is economically feasible is one of the main directions of heating systems development. Autonomous heating systems will be applied where there is no co-generation plant service.

The share of rural consumers is around 30 per cent of the overall heat demand of Kazakhstan. This demand is being covered through burning of various fuels in heating furnaces and small independent heating systems.

Ekibastuz AES (former GRES-1) Thermal Power Plant

Of the overall demand of urban consumers, 43 per cent is provided by co-generation power plants, around 14 per cent by district heat boilers, and 43 per cent by independent heating systems and heating furnaces.

The situation in the district heating sector is the same as for electricity, very limited funds have been spent on maintenance and renovation, and the investment needs are high. The total heat losses in the distribution systems are about 40 per cent.

Almaty City is covered by three District Heating Companies. At present, the district heating tariff in Almaty is approximately $20/Gcal including VAT (0.017 cent/kWh). The tariff for domestic hot water is 200 tenge/m^3 (approximately $1.7/m^3).

Financial environment in energy efficiency and renewable energy

Legal, regulatory and policy framework

The goal and the basic priorities of the development of the electricity sector are presented in the "Programme for the Development of the Electricity Sector up to 2030" (adopted as a special Resolution of the Government of the Republic of Kazakhstan, April 1999). The principal strategic directions of the development of the sector are:
- Creation of an integral power system of Kazakhstan; simultaneous operation with the integral power system of the Russian Federation and the power systems of the Central Asian republics
- Further development of an open competitive power market
- Maximum use of the existing power stations with reconstruction and modernization thereof
- Improvement of the power generation structure by developing technologies using renewable energy resources

- Reconstruction and modernization of the existing heating systems with combined generation of heat and electricity as effective power saving technology, allowing a significant reduction in consumption of fossil fuel and greenhouse gas emissions
- Implementation of modern autonomous high-quality sources of heat, wherever it is reasonable in economic and environmental terms, in comparison to the combined generation of heat and electricity and to the centralized delivery of heat from boiler houses.

The Law on Energy Saving of the Republic of Kazakhstan came into force in 1997, but has not been effective.

New laws on energy saving and on renewable energy are being drafted. To support the development of the Law on Energy Savings, the joint stock company Kazakh Research Institute of Power Engineering named after Sh. Ch. Chokin (KazNII Energetiki) has been contracted to develop a governmental energy saving programme. The programme reviews the economic situation in all sectors, including the energy consumption per produced unit, and presents proposals for items to be included in the law. In June 2008 KazNII Energetiki established the Republican Energy Saving Centre to develop part of this programme.

According to the Ministry of Energy and Mineral Resources, the Law on Energy Savings is expected to be approved by the Parliament during autumn 2009. The Law on Renewable Energy should be adopted during the first half of 2009.

According to the Ministry of Energy and Mineral Resources, the Government is also discussing how to establish a more modern approach to setting the various tariffs. They are also discussing support programmes for low income people that will be the most affected by increased tariffs.

The State Energy Supervision Agency within the Ministry of Energy and Mineral Resources has the operational responsibility for preparing the new Law on Energy Savings. It has also been asked to evaluate and propose the establishment of an energy (efficiency) agency, within or outside the Ministry. International guidance and support would be very welcome.

The Ministry of Environmental Protection has provided input to and comments on the two new laws. A lot of proposals have been provided especially for the Law on Renewable Energy to ensure that all the barriers in this sector could be overcome. There is a high potential for wind farms, but costs are high since several of the potential sites are in rural areas far from the grid. Kazakhstan ratified the Kyoto Protocol on 19 June 2009. In Kazakhstan, substantial reductions of GHGs can be achieved through energy efficiency.

Interest in receiving equity and mezzanine financing

The Chamber of Commerce and Industry has seen an increasing interest from the energy sector. At the beginning of December 2008 they are organizing their first "energy event"; "Kazakhstan's energy sector – realities and perspectives", a round table to discuss the prospects of energy sector development. They expect to see an increasing interest in equity and mezzanine financing in the energy sector.

Energy efficiency investments in the private sector, and especially in industry, seem to be the most promising sector for the time being. According to the KAZSEFF (EBRD) team, this is mainly motivated by the need for upgrading and modernization of production facilities to improve and increase production. New, modern equipment will provide substantial energy savings. KAZSEFF also expressed interest in co-financing of projects with the Fund.

According to the Deputy Major of Almaty, they are interested in partners for the new ESCO that is planned to be established in Almaty City, and would welcome equity participation from the Fund.

The district heating company within Almaty also stated that they very much welcomed participation from the Fund.

Barriers to financing EE and RES

There are several barriers for realization of EE and RES projects:
- Lack of capacities and skills in how to prepare project proposals properly
- Low energy tariffs (electricity and district heating)
- Lack of incentives for energy efficiency investments in the public sector
- Lack of awareness of energy efficiency among municipal decision-makers
- No responsibility for energy efficiency in municipalities
- Almaty: district heating company not able provide enough heat, leading to indoor temperatures lower than norms during the winter season. Because of this, the real savings from various energy efficiency measures will be less, which could be a barrier for ESCO business
- Low rate of energy metering in district heating systems
- Astana: several buildings heated above norms, no interest in installing energy meters
- Electricity from renewable energy sources: no laws and regulations in place securing access to land or to the public grid, no system for feed-in tariffs.

According to the World Bank, the legislation is in place allowing and ensuring private investments in the energy sector, but the low energy tariffs do not attract investors. Due to lack of long term government strategy, the tariffs are also unpredictable. Supported by the World Bank, a Public Private Partnership Centre has recently been set up within the Ministry of Economy and Budget Planning.

Today's market for energy efficiency investments in the public sector seems to be limited due to low tariffs and lack of incentives (reduced costs gives reduced budgets, and cannot be used for repayment of loans, or improved services). In municipalities there are no departments responsible for energy efficiency and there is a lack of awareness of energy efficiency among municipal decision-makers.

TeploKommuneEnergo, covering about 30 per cent of Almaty, is owned 38 per cent by the Akimat City and 62 per cent by the employees. They have 54 heating units, 5.000 pumps, and 300 km of distribution network, but are not able to provide enough heat due to inefficient heat production and high losses in the distribution system. This leads to indoor temperatures lower than norms during the winter season. Because of this, the real savings from various energy efficiency measures will be less, which could be a barrier for ESCO business.

Regarding district heating bills, about 35 per cent is based on measurements, the rest on norms. For domestic hot water, about 75 per cent is based on measurements.

In Astana, a private company had proposed a project with new automatic control systems in 54 schools, with a payback of less than 3 years. The project was not implemented due to the budget code system; savings could not be used to repay the investment (reduced budget next year). Experience from installing energy meters in the residential sector has shown measured energy consumption substantially higher than the norms, mainly due to higher indoor temperatures. Because of this, residents to not want to install energy meters (sometimes installed meters have "easily been broken").

Regarding construction of new buildings, there is no control to ensure that the existing building codes also regulating thermal insulation are being followed.

Regarding electricity from renewable energy sources, there are no laws and regulations in place securing access to land or to the public grid. No system for feed-in tariffs is established, nor discussed. It is still not clear how this will be treated in the new laws on energy savings and renewable energy, but there are some expectations and hopes.

Incentives

Several incentives are expected in the new laws on energy savings and on renewable energy.

It looks as if everyone is waiting for the new laws to be approved and come into force. Meanwhile, there is little progress in the fields of energy efficiency and renewable energy.

Financing schemes

There is no evidence of any national financial schemes for energy efficiency and renewables in operation. Some financial schemes might be proposed in the new law on energy savings and law on renewable energy.

EBRD:
- 2008-2011: KAZSEFF: Kazakhstan Sustainable Energy Financing Facility, provides longer-term loans to enhance the modernization process of the industrial sector and to support the development of renewable energies in Kazakhstan. The Facility consists of two components:
    - A $75 million financing facility for on-lending to industrial enterprises through local Partner Banks;
    - A technical assistance grant to support companies in the identification of energy loss areas, propose technical solutions for lowering energy consumption and prepare bankable projects.

  KAZSEFF is designed for privately owned industrial companies, both medium-sized and larger scale enterprises in Kazakhstan. Under this facility typical energy efficiency investments range from $ 250,000; the maximum loan amount for any one company (for one or several energy projects) should not exceed $ 7 million. The technical and financial experts provide free-of-charge consulting services for investing companies to substantiate the investment proposals to partner banks. If applicable the consulting team conducts energy audits and feasibility studies, gives recommendations on the planned measures and supports the companies in preparing bankable energy efficiency and renewable energy projects.
- The Business Advisory Service (BAS) supports SMEs with management and technical consulting services.
- The Turn-Around-Management programme offers senior advisors for management teams.

UNDP:
- Removing barriers to energy efficiency in municipal heat and hot water supply (2007-2011), including three main components:
    - Strengthening the legal, regulatory and institutional framework to promote energy efficiency of the heat and hot water supply services in Kazakhstan;
    - Enhancing the awareness and building the local capacity to implement and adopt new institutional and financing mechanisms for organizing energy efficient heat and hot water supply services and leveraging financing for them;

- Compiling, analyzing and disseminating the project experiences and lessons learnt and initiating their effective replication in Kazakhstan and in other countries

An ESCO will be established within Almaty City as a part of this project.

- Wind power market development initiative (2004-2008); The objective of the full-scale project is to promote the development of the wind energy market in Kazakhstan by:
    - Assisting the Government to formulate a national programme on wind energy development
    - Providing information for and building the local capacity to develop wind energy products in Kazakhstan and to organize financing them (including site "mapping" and expansion of the wind speed measurement programme)
    - Facilitating construction of the first 5MW wind farm to prepare the ground for and reduce the risks of further investments
    - Monitoring, analysing and disseminating the experiences and lessons learnt during the implementation of the project

Achievement to date:
    - Development process for the 5 MW pilot wind farm construction
    - Wind monitoring programme under preparation for 8 selected sites
    - National wind power programme under preparation
    - Draft law for RES under preparation.

ESCOs

There is no ESCO in operation in Kazakhstan, but an Almaty City ESCO will be established within the UNDP/GEF project on removing barriers to energy efficiency in the municipal heat and hot water supply, and is supposed to be operational in the first quarter of 2009. Almaty City will provide $1 million for a 100 per cent stake in the ESCO. UNDP/GEF will provide $500,000 for energy saving equipment for the first projects.

Banking sector

During the mission there was only one meeting with bank – the Bank TuranAlem, one of the largest banks in Kazakhstan, and the third largest bank in the CIS. A large part of their portfolio is related to investment projects in the energy sector. They have their own staff also doing technical due diligence projects.

Bank TuranAlem confirmed that they are discussing with EBRD to become one of the participating banks within the KAZSEFF programme.

According to Bank TuranAlem, the sector of energy efficiency and renewables is rather new for the Kazakh banking sector, and only few banks are ready to offer financing to this sector at present.

Brief conclusions and recommendations

- There is a large need for new investments in the energy sector due to several years without investments and renovations
- At present, the market for investing in RES does not seem favourable due to low tariffs and lack of legal framework including land concession and access to the public grid

- At present, the market for investing in EE in the public sector does not seem favourable due to low tariffs, lack of awareness to energy efficiency among municipal decision-makers and lack of incentives (savings give reduced budgets)
- EE investments in the private sector, especially industry, seem to be the most promising for the time being. Most of the investments, however, are mainly motivated by the need for more efficient and increased production
- EE has been given high political priority only this year, mentioned by the President in several of his speeches. According to the President, tariffs need to and will be increased
- New laws on energy savings and on renewable energy are being drafted, and are expected to be approved during the next year. The future market development very much depend on the contents of these laws
- There are several barriers to energy efficiency
- If the Almaty ESCO within the GEF/WB project starts operation shortly, this might be an interesting partner for the Fund. Rules for public tendering might be a barrier
- KAZSEFF (EBRD) is interested in co-financing projects with the Fund and in cooperation on capacity building activities
- It would be useful for the Fund Designer and later the Fund Management to learn about the experiences of Bank TuranAlem from financing of the energy sector.

Energy efficiency and renewable energy sources project development and finance capacities

Existing and prospective EE and RES projects

TeploKommuneEnergo (Almaty) has prepared a $ 500 million modernization plan up to 2012, but with no financial solution as of today. The Asian Development Bank can provide a 6 per cent, 14 year loan if government co-financing and guarantee is provided. The company management is not optimistic regarding a financial solution in the nearest future. By next summer, the company expects to have accumulated deficits of 700 mill tenge (6 mill USD).

Within the UNDP project on wind power market development initiative, project documents have been prepared for a 5 MW pilot wind farm construction.

Assessment of investment project development skills

Based on the meetings arranged by the NC and NPI, it is the impression of the consultant that the level of capacities and skills operationally available for the development and financing of energy efficiency and renewable energy investments in Kazakhstan is too low to support a wider market development. Thus there is a need for capacity building.

Public sector

During meetings with the Ministry of Energy and Mineral Resources and the Ministry of Environmental Protection, they stated that capacity building on project preparation would be urgently needed.

Awareness raising and capacity building is also needed at municipal level, including how to prepare local energy efficiency plans.

<u>Private sector</u>

The Republican Energy Saving Centre (established by KazNII Energetiki in June 2008) is the National Participating Institution for the UNECE Project. For them to be able to act as project initiator and evaluator, capacity building is needed.

KazNII Energetiki is the local institution participating in the KAZSEFF team, and they are supposed to be trained on developing bankable projects for the industry. The role and involvement of KazNII Energetiki versus their Republican Energy Saving Centre is not clear, but they will most likely be operating as one unit.

The new ESCO to be established in Almaty will need capacity building to be able to operate professionally.

The District Heating Companies expressed interest in training in how to prepare bankable projects in their facilities.

Within the KAZSEFF project, capacity building will be provided to the local participating banks. EBRD expects 3-6 banks to be involved. This capacity building will take place during the first half of 2009, and the UNECE project should be able to benefit from this.

<u>NGOs</u>

The Centre for Energy Efficiency and Cleaner Production in Almaty, established within the Norwegian funded capacity building programme, has experience from developing more than 100 energy audits in the building sector, and from implementing some pilot and demonstration projects. It would be beneficial for them to be involved in preparing projects for the Almaty ESCO and capacity building in preparing business plans.

Assessment of equity and mezzanine financing business development skills

As we did not learn about any large scale RES project implemented or any scale special purpose companies established, we expect the experience, knowledge and skills on equity and mezzanine finance to be very limited.

Brief conclusions and recommendations

- There is a substantial need for capacity building in Kazakhstan, both related to investment project development and equity and mezzanine finance business development. There are some experts who seem to have capacities and skills that could be utilized for the new Fund, but the resources are too limited to support a wider market development;
- Guidance, support and capacity building is also needed for involved ministries.

Investor interest

Public sector

The issue of potential participation in the Investment Fund was presented and briefly discussed with some parties during the mission. No public sector interest can be reported. However, it is important to bear in mind that we did not meet with many real candidates for such a potential interest.

Private sector

The issue of potential participation in the Fund was presented and briefly discussed with the Bank TuranAlem, one of the largest banks in Kazakhstan, and the third largest bank in the Commonwealth of Independent States. The Chairman of the Board is a former energy Minister of Kazakhstan. A large part of their portfolio is related to investment projects in the energy sector. They have their own staff also doing technical due diligence of projects.

Bank TuranAlem expressed interest in receiving more information to evaluate a possible participation in the Fund. They seem to have experience that would be of interest to the Fund Designer and later the Fund Management.

The interest from international financial institutions needs to be clarified by their headquarters, and not through the branch offices in Kazakhstan.

Potential partners for project co-financing

If not deciding to participate as investor in the Fund, Bank TuranAlem is interested in co financing projects.

KAZSEFF (EBRD) also expressed interest in co-financing of projects with the Fund.

Brief conclusions and recommendations

Regarding the Eastern European Energy Efficiency Investment Fund, there were meetings with few potential public and private investors in Kazakhstan. Bank TuranAlem will consider participation in the Fund. The interest from international financial institutions needs to be clarified by their headquarters, and not through the branch offices in Kazakhstan.

REPUBLIC OF MOLDOVA

Energy overview

Existing energy resources, energy dependence for primary and secondary energy resources, production of electricity and of heat, use of renewable energy sources (RES)

The period following independence in 1991 has been characterized by a significant decrease in energy use in the country. This reduction in energy consumption is not a sign of improved energy efficiency, but is due to many other factors inherent in the transitional period, including production crisis, financial difficulties and irregular energy supply. Final energy consumption in 2005 was still 65 per cent lower than in 1990.

Despite this large decrease, the economy of the Republic of Moldova still has a high level of primary energy consumption per unit of GDP compared to averages for countries in the Organization for Economic Co-operation and Development (OECD). According to IEA, in 2005, the energy intensity of Moldova (energy use compared to GDP at purchasing power parity (PPP)) is 0.45 toe/ $1000, nearly three times higher than the EU-27 average.

Significant changes have occurred in the fuel balance of the power industry. Coal consumption has substantially decreased, while natural gas has become the main fuel for the power stations and boiler houses and has reached a share of 69 per cent of total primary energy supply (TPES).

Moldova has insignificant reserves of solid fuels, petroleum and gas, and a low hydroelectric potential. This leads to a fuel balance with a high dependence on energy imports (mainly from the Russian Federation and Ukraine) – with import levels reaching 98 per cent of total consumption.

In total, about 75 per cent of the urban dwellings in Moldova are supplied by district heating systems. In the early to mid-1990s heat supplies were still adequate although inefficient, while from the mid-1990s a progressive collapse of the heating networks throughout the country started. The heating infrastructure in the capital of Chisinau is in serious disrepair. Non-payment is chronic. The two biggest district heating systems have very long distribution lines, which contributes to increased losses. In many parts of the city the district heating system is simply being dismantled, replaced by electric heaters or gas boilers in the buildings.

The residential sector is the greatest energy consumer (40 per cent of total final consumption), followed by industry (21 per cent) and transport sector (15 per cent). Agriculture, although dominating in the economy of the country, has a small share in the final consumption of commercial energies (4 per cent).

The economic and structural reforms in the country resulted in a substantial reduction of industrial production, which in turn resulted in reduced energy consumption. However, the energy efficiency of the industrial sector is low. The specific energy consumption in processes is high and the energy losses are substantial. Both energy audits and implemented energy efficiency projects demonstrate high energy efficiency potential in all sub-sectors of industry. Nonetheless, energy efficiency is still not a matter of great concern in industry.

Level of priority given to EE and to RES in country's energy policy

After several institutional reforms, energy is now under the responsibility of the Ministry of Economy and Trade. It is also dealing with technological aspects of energy efficiency. The Ministry of Ecology and Natural Resources (MENR) focuses on environmental aspects of EE and promotes energy efficiency through GHG reduction.

Although primary legislation on energy efficiency and renewable energy exists in Moldova (Law on Energy Conservation (2000) and Law on Renewable Energy Sources (2007), secondary legislation to make it operational is not yet developed. Furthermore, current energy efficiency legislation needs harmonization of laws in terms of intent and objectives. The economic incentives to stimulate the implementation of the recommended measures have to be established. Neither payment for energy nor assessment of pollution payments represent real incentives for energy efficiency increase and renewable energy sources use. Current methodology to establish the cost for energy does not offer advantages to switch to renewable energy use.

Biomass heating system (straw bale boilers) installed at a school in Taraclia village, Causeni raion.

Investments in the energy sector

The Republic of Moldova has insignificant reserves of solid fuels, petroleum and gas, and a low hydroelectric potential. Investments in the energy sector are low.

The Energy Strategy (2007) of the Republic of Moldova up to 2020 has three strategic objectives: (i) security of energy supply; (ii) promoting energy and economic efficiency; and (iii) liberalization of the energy market and restructuring of power industry. The document elaborates on energy saving and enhancing energy efficiency. It indicates key principles that shall guide policy and normative formulation, specific objectives in the field of increased energy efficiency and recommended measures to implement in order to achieve such objectives. The Energy Strategy also dedicates a section to international cooperation in which it recognizes the critical importance of and calls upon international development assistance in achieving its objectives.

The National Programme (2003) of Energy Conservation for 2003-2010 sets out quantitative targets for efficiency improvements at the national level, priority areas for energy conservation and indicates activities to carry out in order to achieve stated objectives. The Programme identifies the priority actions and aims at increasing energy efficiency by decreasing energy intensity by 2-3 per cent annually. It sets an objective to substitute about 6 per cent of the total energy supply with local and renewable energy sources by 2010.

However, for both these documents it is not clear what financial resources are available and what the status of their implementation is.

There are expectations that investments may be attracted from the following sources:

- In the subsectors that represent natural monopoly (electricity transmission, natural gas transportation), investments will come mainly from the companies involved, with possible support from the state budget. For example, in 2008 the actual combined investments by the distribution companies (JSC "RED Union Fenosa", JSC "RED Nord" and JSC "RED Nord-Vest") were approximately Lei 325 million (about €23 million). Planned investments for 2009 are 377 million (about €27 million). Investments in this sector are also possible through loans from IFIs, if returns on the investment can be guaranteed;
- In the sectors that have been liberalized (e.g. electricity supply and distribution, oil market and potentially heat supply), investments may come from private investors if a favourable investment climate is in place and the investor has a reasonable guarantee of returns on his investment. For example, in December 2008 the Government signed an agreement with the Czech company United Energy Moldova (UEM) on building a new coal-powered thermal power plant with installed capacity of 350 MW at Ungheni. UEM plans to invest €600 million in this project.

A number of measures to attract investments in the energy sector are listed in the Energy Strategy:

- Promoting legal reforms that facilitate project finance and attractive investment climate
- Establishing and strengthening mechanisms for attraction and efficient use of financial resources for financing of energy projects
- Implementation of innovative financing mechanisms, based on the settlement of external debts by offsetting part of the debt with equivalent state investments in environmental protection or development of renewable energy sources
- Promoting private investment in CDM projects and, eventually, entering into appropriate agreements to work within the EU Emission Trading Scheme
- Use of internationally approved methodologies for estimating the amount of required investments to reach strategic objectives and specific objectives for each segment of the country's energy industry and for prioritizing development programmes
- Developing an information database for energy projects financing
- Supporting reforms in the banking sector.

Financial environment in energy efficiency and renewable energy

Legal, regulatory and policy framework

The national energy policy is still focused more on the industrial (21 per cent of the total final energy consumption in Moldova) than the residential sector (40 per cent of the final energy consumption), and lacks the follow-through of the development of realistic action planning and programme implementation. The existing normative-legislative framework must be extended, especially concerning: (i) imposing some energy and environment efficiency requirements (standards); (ii) elaboration of national programmes and strategy for energy efficiency, firstly in the housing and public sector. Some important regulations under development should therefore be adopted in a more functional form in order for the energy conservation policy to succeed (law on ESCO formation; regulation regarding energy conservation incentives etc).

Strong points of the current energy efficiency legislative framework:
- Energy efficiency is a priority in the Republic of Moldova and strategic policy objectives for energy conservation are defined. Important laws are in force: Law on energy conservation, Law on electricity, Law on natural gas, Law on renewable energy sources
- The national objective of an annual 2-3 per cent decrease in the energy intensity of GDP, stipulated by the Energy Strategy, is a very ambitious task
- There is political will to improve current EE and RES legislation.

Weak points of the energy efficiency legislative framework:
- Energy efficiency legislation is more declaratory than operational;
- Actions in the area of developing and implementing secondary legislation, institutional capacity building, developing sectoral programmes, and securing financing are required.
- Discrepancies and contradictions between laws;
- While legislation considered incentives for energy efficiency activity development, the lack of mechanisms and underpinning legislative framework stand as the most significant barriers against their implementation.
- Inefficient legislation to support the structure of the Agency for Energy Efficiency and Renewable Energy Sources (former National Agency for Energy Conservation) under the Ministry of Economy and Trade and lack of financial and human resources for its operation.

National priority areas for EE and RES

Energy efficiency

Energy efficiency is one of the priorities for the national economy and for the energy sector and has been named a key objective under the EU-Moldova European Neighbourhood Policy Action Plan (Objective 66).

Specific national objectives in the field of energy saving and increasing energy efficiency include:
- Implementation of the National Programme for Energy Conservation 2003-2010 and its subsequent extension
- Developing, approving and applying standards aimed to increase efficiency of energy consuming equipment, in line with standards set out in the EU legislation on energy efficiency
- Raising awareness of energy saving initiatives that increase energy efficiency in the public sector (organizations funded from the state budget), households and in the industry and energy sectors
- Promoting the use of efficient, economically viable and non-polluting energy technologies and equipment in all sectors of the national economy
- Encouraging the application of new rules for investments and incentives to increase energy efficiency
- Establishing a database on energy efficiency options and providing free access for legal entities and individuals to this information
- Promotion of consultancy and auditing services by private and state organizations, which will provide information about energy efficiency programmes and technologies, as well as technical assistance to public and private sector consumers

- Setting up regional energy efficiency demonstration centres
- Developing pricing and taxation policies which would provide clear signals favouring energy efficiency.

Renewable energy resources

The Republic of Moldova has set a goal of increasing the share of RES in the country's energy balance up to 6 per cent by 2010 and up to 20 per cent by 2020. A document outlining the current state, key tasks and means for development of renewable energy sources in the medium term has been developed under the title Draft National Programme for the Development of Renewable Energy Sources up to 2010. This Programme has not been approved but its main elements have been incorporated into the Energy Strategy of the Republic of Moldova up to 2020.

The emphasis in developing the potential of renewable energy is on the following RES:
- Biomass: use of biomass both from plants grown specifically for production of biofuel and from agricultural, timber and urban waste
- Solar energy (conversion to electricity and heat)
- Wind energy
- Hydropower

Specific national objectives in the field of renewable energy sources include:
- Developing and improving legislation on the renewable energy sources. The Republic of Moldova intends to transpose into the national legislation Directive 2001/77/EC on promotion of electricity produced from renewable energy sources in the internal electricity market and Directive 2003/30/EC on promotion of the use of biofuels or other renewable fuels for transport
- Establishing conditions for the stable development of the energy industry based on RES and increasing the amount of RES use in the national economy, including identification and removal of barriers to utilization of RES
- Increasing the level of professional training of personnel in this field
- Increasing public awareness of the importance of RES use for the sustainable development of the country
- Establishing a national fund for promoting renewable energy.

Interest in receiving equity and mezzanine financing

Capital markets on the whole are underdeveloped and shallow. The primary market for government securities is well-organized but dominated by commercial banks, while the secondary market is almost non-existent. Liquidity on the markets is constrained by the absence of non-residents, who have sold their holdings in the wake of the 1998 Russian crisis and have not returned since then. The lack of investors on the equity market is exacerbated by fragmentation of the market.

Barriers to financing EE and RES

All sectors have big energy efficiency potentials in the building stock, technologies and management. Lack of financing and insufficient or non-existent economic incentives are major barriers for the implementation of energy efficiency measures by all types of consumers.

<u>Residential</u>

The residential sector is characterized by a broad range of inefficiencies starting from large commercial heat and hot water losses (thefts and leakages) in the systems, to wasteful end use in dwellings that are not weatherized. However, energy efficiency projects in the residential sector are not implemented with the exception of donor-assisted demonstration projects or occasional attempts by managers of some buildings.

Meeting at the Institute of Power Engineering, Academy of Sciences

The increase in energy price levels, combined with the low level of income (the energy bill of a household can represent more than 50 per cent of an average salary), has resulted in low rates of payment collection (still, the payment of electricity bills increased after the Law on Energy allowed disconnection of non-paying consumers). Actually, it is difficult to reach real savings in the residential sector under actual conditions, where the energy comfort is much lower than the norm, combined with difficulties in defining and explaining the "baseline energy consumption".

Increase in prices of imported energy fuels, big system losses, and chronic non-payment by consumers, led to a serious financial crisis of the heating sector. Since 2000, heating companies have been established as municipal enterprises, wholly owned by the local administration. The majority of these units did not have the necessary experience and financial means to redress the problems of the heating systems, or at least for reducing the crisis. In fact, many heat supply systems have stopped operation, whereas the efficiency of the remaining ones is very low. As a result, heating services both to public and residential buildings continue to deteriorate. Many customers have started refusing the heat supply, thus reducing significantly the demand for heat, consequently affecting the efficiency of generation.

Finally, an important barrier for the implementation of energy efficiency improvements in multi-story buildings of the residential sector is the lack of legal authority of the housing associations. As there is a high share of private ownership, the associations cannot oblige the individual owners or tenants to participate in the funding of energy efficiency measures. The privatization of apartments has left the apartment owners without any obligations regarding common facilities such as heat supply, maintenance of the building shell, etc.

Public sector

The legislation does not give incentives to local governments and public institutions to save energy. This, according to mayors, principals, directors of public institutions, is the main impediment why energy efficiency projects are not implemented in these given institutions.

The current budgetary regulations should be improved to give more incentives to local councils to manage their budgets rationally. The limits imposed by the central authorities on the budget expenses do not encourage local governments to develop and implement different projects that would result in reduction of costs for municipal services, including heat. Under the current itemized budget allocation system, municipal budget saving in any expenditure category will result in a reduced allocation from the central budget in the following year. For example, if a municipality implemented an energy efficiency project in schools and thus reduced the fuel consumption by 20 per cent, the following year the central authority would cut that 20 per cent (money equivalent) from the local budget and local budget would get no savings out of the project and would not use savings for alternative needs in the public entity (for example buy new books for the school) or to improve the quality of services in public buildings, etc.

Budget constraints, together with the deterioration of the district heating services, have resulted in great difficulties for assuring enough heat for the public institutions such as schools, kindergartens, hospitals, etc. Mayors of municipalities have addressed the problem of budget creation but no real improvements have been carried out up to now.

Industry

EBRD experience shows difficulties in finding customers in the corporate sector reaching the minimum requirements in regard to international banking practices. Furthermore, lack of financing leads to a concentration on core business activity and little interest for energy efficiency investment opportunities. Although energy audits and implemented energy efficiency projects demonstrate high energy efficiency potential in all subsectors of industry, energy efficiency is still not a matter of great concern in industry. The limited awareness of the industrials about the possible benefits and needed actions results in lack of interest and motivation.

Incentives

Economic incentives to stimulate the implementation of the energy efficiency measures have to be established.

Financing schemes

So far projects in the area of energy efficiency have been financed mainly by international financial institutions and/or international donors.

The World Bank has provided a $10 million loan for investments in the thermal power/district-heating sector as part of the bigger loan (Energy 2) for energy projects. The loan is to be used in the first stage for investments in four cities (Cantimir, Falesti, Straseni, and Floresti). A pilot project is underway in Ungheni in which four small heating plants are being constructed. Under the conditions of this loan only heating for institutional buildings can be financed. However in the case of Ungheni, the capacity of the heating plants constructed is large enough to accommodate local residential demand, if funding can be found to connect residential buildings to the heat distribution network.

EBRD is working of the implementation of a credit line dedicated to energy efficiency in Moldova, based on the experience of similar EE credit lines in Bulgaria and Ukraine. EBRD has been conducting a study on the EE market in Moldova to be completed by the end of 2008. EBRD could in principle consider participation in the Investment Fund to be established but the decision on this would be taken at the EBRD Headquarters rather than in the country office.

USAID has completed four projects related to energy efficiency: Energy sector regulatory development, municipal network for energy efficiency, Moldova energy efficiency weatherization, and power sector privatization. However, three years ago USAID decided to phase out energy projects and there are no current or prospective ones in this area.

The EU TACIS Programme has funded energy efficiency projects, most notably a project involving 40 industrial energy audits; the creation of the National Energy Conservation Agency (ANCE); financing for energy auditing equipment for ANCE; and implementation of the recommended measures at one or two of the audited enterprises. This project had a capacity building element (intended to build capacity for energy auditing at the ANCE) but this capacity has to a large extent been lost due to the lack of follow-up, staff movements and eventual transformation of ANCE into a new Agency on Energy Efficiency, which was not yet operational at the time of the assessment mission.

The Norwegian Ministry of Foreign Affairs is financing a 4-5 year project on cleaner production and energy efficiency in Moldova. In the framework of the project a small revolving fund for cleaner production and energy efficiency ($30,000) has been established.

ESCOs

Lack of financing has limited the demand for energy services until now. No real ESCOs have been created but there are engineering companies that have worked on donor-financed turn-key contracts in the range of $50,000 to $150,000. One of the main barriers for energy services development is that energy prices are too low, which does not allow energy savings as a result of energy efficiency projects to compete with other types of investments. Legislative impediments in the public sector, such as budgetary proceedings, represent another barrier for energy efficiency projects to be implemented.

Banking sector

Moldova performs above average in the use of international accounting standards – in 2002 almost 82 per cent of the sampled companies were using them, and about 46 per cent had external auditors. This should allow banks and other potential lenders to more easily assess credit worthiness.

Moldovan companies perceive themselves to be credit constrained, but their difficulty in accessing credit is similar to other transition economies. Long-term bank lending remains limited, with few companies being able to obtain loans with terms longer than 12 months. Nominal and real interest rates and collateral requirements are also high.

In 2005 new investments in the corporate sector were mostly financed on internal funds. The median debt-to-asset ratio is still very low, suggesting that companies are not becoming more leveraged and that access to affordable finance is still problematic.

While being overwhelmingly dominant in the financial sector, the banking sector is still relatively small. The relatively low level of financial intermediation is partially due to the early stage of market development and the widespread use of cash in the economy. Despite the continuous growth of deposits in banks indicating the increase of confidence in the banking sector, cash holdings are still prevalent and are supported also by the inflow of remittances from abroad through official and non-official channels.

Lack of product diversification, poor banking skills, and inadequate lending and crediting policies are de facto limiting the efficient development of new banking services. The costs of these inefficiencies are eventually transferred to the private sector as high interest margins and transaction costs.

In addition, weaknesses in the operational environment are undermining the efficiency of the banking sector. Uncertainties in policy implementation by the authorities complicate bank operations, and increase costs. Since monetary operations of the National Bank of Moldova are often unpredictable for banks, they tend to keep higher liquidity. The recent increase in the level of required reserves raises the cost of compliance, exacerbating existing inefficiencies in the system.

Brief conclusions and recommendations

- Although primary legislation on energy efficiency and renewable energy exists in Moldova, there is lack of secondary legislation. Thus energy efficiency legislation is more declaratory than operational.
- Economic incentives to stimulate the implementation of the recommended measures have to be developed.
- Inefficient legislation to support the structure of ANCE. Most important barriers posed to the fulfilment of Agency activity are the lack of financial and human resources.
- All sectors have big energy efficiency potentials in the building stock, technologies and management. However, lack of financing is a general barrier for the implementation of energy efficiency measures by all types of consumers.
- According to 2005 data, new investments in the corporate sector have been mostly financed using internal funds. Capital markets on the whole are underdeveloped.

Energy efficiency and renewable energy resources project development and finance capacities

Existing and prospective EE and RES projects

The Republic of Moldova implements activities related to the Clean Development Mechanisms under the Kyoto Protocol. Moldova ratified the UNFCCC in 1995 and the Kyoto Protocol in 2003. Being a non-Annex I party, Moldova is eligible for CDM projects. Five have already been implemented in Moldova and several more are under preparation. Future UNECE project activities could be related to these CDM projects activities.

See information on other EE projects in section under *Financing schemes*.

Assessment of investment project development skills

Public sector

The assessment mission showed the need for capacity reinforcement on EE and RES issues.

At the institutional level, the lack of secondary legislation on EE and RES is a large barrier for project development. Though the experts encountered in the Ministry of Ecology and Natural Resources showed good understanding of the issues, financial and human resources to deal with the issue are insufficient. The Ministry of Energy has been dissolved and its tasks have been transferred to the Ministry of Economy and Trade, which led to a discontinuity in implementation of the energy-related activities. The same could be said regarding transformation of the National Agency on Energy Conservation into the Agency for Energy Efficiency and Renewable Energy Sources under the Ministry of Economy and Trade. Institutional stability is required and the existing normative and legislative framework must be extended.

Private sector

The lack of financing for core business activity limits the interest for energy efficiency investment. The limited awareness of the corporate sector of the possible benefits and needed actions results in little activity in energy efficiency investments.

Furthermore, market conditions have not been favourable for the development of ESCOs in Moldova until now, one of the main reasons being the low end-user energy prices, which did not allow energy savings to compete with other types of investments. Since the conditions in the market are changing, there are premises for ESCO development. The recent rise in energy prices increased the share of energy costs in the industry, which made it more important to the managers of the respective entities. New energy efficient technologies are increasingly available on the market, which are offered using financial instruments – for example leasing.

Although there is a strong demand for leasing vehicles, especially from SMEs, the leasing industry is still underdeveloped even by regional standards mainly because of tax disadvantages for those who lease equipment. Development of leasing mechanisms could be facilitated if the VAT status was adjusted appropriately.

NGOs and academic organizations

Several organizations involved in activities related to energy efficiency and renewable energy sources were visited and interviewed at the time of the assessment mission.

The Institute for Development and Social Initiatives (IDIS "Viitorul") is an independent think-tank, a member of several national and international networks of public policies. Its main fields of economic research include energy policy. IDIS "Viitorul" recently made an assessment of public policy in the gas sector and provided support to Chisinau Municipality in the sustainable provision of heating services. IDIS "Viitorul" has a good knowledge of the Moldovan banking sector. Under the EBRD project on establishing energy efficiency credit line, IDIS "Viitorul" recently assessed the Moldovan banks practices and capacity-reinforcement needs.

ProEnRe is an NGO working on the promotion of RES in Moldova. Closely related to the Technical University of Moldova, it conducts research studies on RES, lobbying and also conduct prefeasibility studies on solar and wind energy projects. It has created databases of RES potentials and methodologies that could be consulted during future ECE project activities.

The Institute of Power Engineering of the Academy of Sciences of Moldova, the Agency for Innovation and Technology Transfer of the Academy of Sciences of Moldova, and the Technical University of Moldova have strong scientific and technological capacity for developing and implementing EE and RES projects but so far this has only been done on a small scale and as pilot projects.

All sectors have big energy efficiency potentials in the building stock, technologies and management. However, lack of financing is a general barrier for the implementation of energy efficiency measures by all types of consumers. Equity and mezzanine financing business development skills are therefore limited.

Brief conclusions and recommendations

- Technological skills are reasonably well developed in the Republic of Moldova, both on energy efficiency and renewable energy issues. However the assessment mission showed the need for capacity reinforcement on business development issues.
- Though the experts encountered showed good understanding and capacity on EE and RES issues, financial and human resources are insufficient to implement practical actions. Institutional stability is required and the existing normative and legislative framework must be extended.
- The assessment underlined that lack of financing for core business activity limits the interest for energy efficiency investment. The limited awareness of the corporate sector of the possible benefits and needed actions results in little activity in energy efficiency investments. Equity and mezzanine financing business development skills are therefore limited.
- The market conditions have not so far been favourable for the development of ESCOs in Moldova.

Investor interest

There is strong interest in implementing energy efficiency projects and developing new ones from a number of institutions in the country. However, neither public nor private entities in the country have expressed particular interest in participating as potential investors.

Public sector

The Ministry of Economy and Trade, as well as the Ministry of Ecology and Natural Resources welcomed the UNECE initiative. However, there was no commitment for potential participation of the Government of the Republic of Moldova as an investor in the Investment Fund or in co-financing potential projects in Moldova.

Private sector

The Moldovan Chamber of Commerce and Industry (MCCI) is a non-governmental, autonomous and independent organization which represents the concerns of its members to governmental authorities and foreign business circles. MCCI welcomed the UNECE project and is ready to be involved in further project developments. Among its missions, MCCI provides information and training to its members. MCCI also has a very good relationship with the banking sector.

Potential partners for project co-financing

No potential partner for project co-financing could be identified at this stage. However, the Environmental Fund (under the Ministry of Ecology and Natural Resources) and Energy Efficiency Fund (under MET but not yet functioning) may be potential sources of co-financing. In addition, EBRD may be interested in becoming an investor in the Investment

Fund or be involved in co-financing of some projects. These decisions will not be taken at the country office level but at EBRD Headquarters.

Brief conclusions and recommendations

- While there is no strong interest among the public and private entities in the country to become investors in the Investment Fund, certain possibilities remain for co-financing of specific projects.
- EBRD may be interested in becoming an investor but the decision on this will have to be taken by EBRD Headquarters.

ROMANIA[7]

Energy overview

Existing energy resources, energy dependence for primary and secondary energy resources, production of electricity and of heat, use of renewable energy sources (RES)

The period following 1990 has been characterized by a significant decrease in energy use in the country. This reduction in energy consumption is not a sign of improved energy efficiency, but is due to many other factors inherent to the transitional period. Final energy consumption in 2005 was still 40 per cent lower than in 1990.

Despite this large decrease, Romania's economy still has a high level of primary energy consumption per unit of GDP compared to averages for OECD countries. According to IEA, in 2005, the energy intensity of Romania is 0,22 toe/$ 1000 at ppp, 38 per cent higher than the EU-27 average.

Romania has significant reserves of solid fuels, petroleum and gas. It has traditionally been an oil producing country but, as the reserves have been declining during the past decade, Romania has become a net importer. The production of natural gas also faces a certain decrease while importations from the Russian Federation are rising sharply. Coal (mostly lignite) production is increasing, in physical units as well as in weight in the total production. Firewood and agricultural waste still hold a relatively significant share in the internal production of primary energy (around 9 per cent of TPES). They are especially used in the rural areas by conventional technologies.

According to IEA, in 2005, electricity is mainly produced from thermal power plants (37 per cent of the total electricity production is coal fired and 16 per cent gas fired, cogeneration being well developed in Romania), hydropower (34 per cent) and nuclear (9 per cent). Indeed, in the late 1970s Romania engaged a nuclear programme aiming to build five 700MW nuclear units. The first unit started producing in 1996 and the second unit in 2007. It is estimated that units 3 & 4 will be completed by 2012 and 2025 respectively, financed through a public-private partnership.

Approximately 30 per cent of Romania's total building stock receives its heat and hot water from district heating systems, a figure that rises to 58 per cent in urban areas. District heating accounts for about 31 per cent of the country's total heat and hot water demand. In the residential sector, in 2003, more than half of the population obtained its heat supply by burning solid fuel (coal, firewood) in stoves, mainly in the rural areas, but also in many cities.

Industry is the greatest energy consumer (33 per cent of total final consumption), followed by the residential (31 per cent) and transport (17 per cent) sectors. While energy consumption is decreasing slightly in industry, it is increasing in the residential and transport sectors.

Industry consumes mostly fossil fuels (70 per cent of the total final consumption of the sector). The chemical and petrochemical, metallurgical and construction sectors are the principal industrial consumers. In 2003 the National Agency for Energy Conservation (ARCE) estimated the potential for energy conservation at around 17 per cent of the sector consumption.

Residential energy use has been increasing since 2000. According to ARCE, thermal energy consumed in housing represents 82 per cent of the total sector consumption (57 per cent for heating and 25 per cent for hot water). The main energy sources are natural gas and

---

[7] No mission was held in Romania. The assessment is based on existing documentation.

district heating (both 35 per cent of the final energy consumption). Electricity represents 9 per cent of the final energy consumption.

During the last years the process of harmonization of the energy efficiency framework of Romania with European legislation has been a priority. Energy efficiency activity is a matter of high priority at national level and increased attention is given by national authorities to energy efficiency issues.

Investments in the energy sector

Romania has significant reserves of solid fuels, petroleum and gas. It has traditionally been an oil producing country but, as the reserves have been declining during the past decade, Romania has become a net importer. The investments in the energy sector have been estimated at more than 8 000 M€ for the period 2006-2009 (Electricity: €3,600 million; gas: €1,600 million, petrol: €1,350 million, mining €1,000 million and renewables €500 million).

In the late 1970s Romania engaged a nuclear programme aiming to build five 700MW nuclear units. The first unit started producing in 1996 and the second unit in 2007. It is estimated that units 3 & 4 will be completed by 2012 and 2025 respectively, financed through a public-private partnership.

The national strategy concerning the modernization of district heating systems developed by the Ministry of Public Administration and Internal Affairs identified a need for annual investments of $450 million over the period 2002-2016. The funds will be raised by bank loans, public-private partnerships, EU grants as well as funds from the State and local budgets.

Financial environment in energy efficiency and renewable energy

Legal, regulatory and policy framework

The institutional framework for the promotion of measures to encourage the efficient use of energy was created in 1990 with the founding of the Romanian Agency for Energy Conservation (ARCE). The powers of this agency were strengthened in 2000 with the adoption of Law 199/2000 regarding the efficient use of energy, this law being amended and supplemented by Law 56/2006. Legally, ARCE is considered to be the main specialized body, at national level, in the field of energy efficiency. It is subordinate to the Ministry of Economy and Finance. The energy efficiency policy of the Government is mainly focused on the residential and industrial sectors so far. Sectoral energy efficiency policies for services and transport are still not developed.

The National Strategy for Energy Efficiency, which is the most important document concerning the energy efficiency policy in Romania, was approved by the Government in 2004 and the Action Plan included is being implemented. An Inter-Ministry Working Group approved by the Government is the body designated to coordinate the implementation of the National Energy Efficiency Strategy and the Action Plan.

Interest in receiving equity and mezzanine financing

At the company level, the credit available to the enterprise sector is mostly debt finance (nearly all bank loans, but also corporate bonds in selected cases). However, according to EBRD, private equity providers are able to find investment opportunities and to successfully exit from them. However, their track record is short and therefore it is difficult for them to raise private funds. Availability of equity for local medium-sized companies is virtually non-existent.

Private equity funds were successfully encouraged to invest in the region from the late 1990s, including donor-supported funds like the Romanian Post Privatization Fund or the Black Sea Fund. However, in the past the level of private equity investments in Romania has been below that of other new EU member states mainly due to delays in foreign investment and the lack of deal opportunities caused by the country's slow economic reform prior to 2000.

The evaluation of the post-privatization funds rated the performance of the Romanian framework as good and highlighted a number of key lessons learned. Firstly, given that the transaction costs of equity investments are high, it is important to ensure that smaller investments are financially viable. Secondly, a significant proportion of potential private equity deals do not come to fruition because incumbent shareholders/managers are not ready to enter into the close partnership – entailing the sharing of information, decision-making and rewards – on which the concept is based.

The country is now viewed as holding some of the region's greatest potential for private equity and has attracted substantial interest from private equity funds. Many regional funds maintain an office in Bucharest (including Mideuropa Partners, Advent, Southeast Europe Fund, GED etc.). At the end of 2006, private equity investments in Romania represented 0.1 percent of GDP, ahead of the Russian Federation, Croatia and Ukraine, but below the Czech Republic (0.4 percent), Hungary (0.6 percent), Poland, Estonia, Kazakhstan and Bulgaria.

Barriers to financing EE and RES

Romania's economic performance has been remarkable in recent years, although important vulnerabilities remain. Romania steadily converges in income, competitiveness, and living standards towards the EU, but the gap remains large. In 2007, income per capita was around 40 per cent of the EU 27 average. Since 2000, the Government has implemented macroeconomic and structural policies which are supportive of growth and disinflation. In the World Bank Doing Business 2007 report, Romania was ranked as the top reformer in Europe and the second in the world for the period 2005-2006. Romania is now a visible and attractive destination for international investors as a result of EU membership, better sovereign ratings and improved access to international capital markets. FDI inflows are estimated at around 6 per cent of GDP in 2007.

Industry

Romanian industry, facing for a long time a deep restructuring process, particularly in the field of mining and quarrying, is still tributary to raw materials imports, influenced at the same time by the lack of investments for production modernization and recovery. However, during 2005-2007, industry recorded an upward trend, mainly in the manufacturing sectors. In 2007 the Romanian industry sector accounted for around 26 per cent of GDP.

Energy efficiency in industry has been promoted since 1991, with important involvement of ARCE and of international donors. The targets are the high energy consuming branches, respectively the chemical and petrochemical industry, the steel and metallurgic industry, the machine building industry and the pulp and paper industry. Nevertheless, industry still has great potential for energy savings. According to Romania's National Strategy for Energy Efficiency, energy savings were estimated in 2001 to 13 per cent of final consumption.

The existent house building stock in Romania comprises approximately 8 million dwellings in 4.6 million buildings. Over 98 per cent of the residential building stock in Romania is privately owned. According to ARCE, thermal energy consumed in housing represents 82 per cent of the total consumption (57 per cent for heating and 25 per cent for hot water). Within the National Strategy for Energy Efficiency the residential sector is given high importance because of the very high potential for energy conservation (35 to 50 per cent of the sector total consumption). These savings are to be obtained mainly by applying important measures for the thermal rehabilitation of the building envelopes, an increase in energy efficiency of the heating systems and also of the lighting systems. Starting from 2007, an energy efficiency certificate is issued for all newly built buildings and, beginning in 2010, an energy efficiency certificate will be issued for single-family dwellings and for apartments in existing residential structures that are sold or rented.

Although the Condominium Law requires households in multi-apartment buildings to organize into Housing Associations and enables them to take investment decisions based on the majority of votes, the mix of income groups living in the same building makes the decision process difficult, given that low-income families are unable to afford these investments. To tackle the affordability issue, a well-designed subsidy system targeted at low-income households would be required in order that these investment programmes can go ahead.

Public sector

In the municipal sector, regulatory reform is progressing across the various sub-sectors (water, waste-water, urban transport, solid waste, district heating). According to EBRD, the environmental upgrades required to meet EU standards are estimated at €30 billion during 2004-2018, a large proportion of which is to be borne by the municipalities and municipally-owned utilities. The availability of co-financing on a non-guaranteed basis remains uncertain as project preparation; utilization capacity and co-financing ability are generally weak at the municipal level. These issues are particularly apparent in smaller municipalities, which to date have not had the benefit of either IFI or local bank financing and lack experience in this area.

Approximately 30 per cent of Romania's total building stock receive their heat and hot water from district heating systems, a figure that rises to 58 per cent in urban areas. District heating accounts for about 31 per cent of the country's total heat and hot water demand. While the rehabilitation of the infrastructure for heat production, transport and distribution is certainly of major importance, measures directed at improved metering and control of heat supplies to final consumers are also crucial. Several municipal heat suppliers are working on the improvement of this situation, often in cooperation with international donors like EIB and EBRD.

The project DHCAN – District Heating and Cooling/CHP in the framework of the SAVE/ALTENER Programme – analysed the current status of district heating systems and collected data in 50 municipalities with the purpose of creating a database of possible investment projects targeting the rehabilitation of district heating systems, mainly through the promotion of small and medium scale cogeneration capacities. A guide for local authorities was also elaborated entitled "Guide for district heating systems modernization through the utilization of small and medium size CHP".

The Law defines obligations for the energy users with an annual consumption above 1000 toe/year (>510 companies in 2006) and of municipalities with more than 20,000 inhabitants (>100 localities in 2006) to accomplish an energy audit every year and develop energy conservation programmes. Consumers with consumption above 200toe/year are obliged to accomplish an energy audit every two years. The energy audit must be executed also by the administrative responsible of public buildings with total unfolded surface larger than 1,000 m², to be renewed every 5 years. The Law also stipulates the elaboration and introduction of energy efficiency standards for energy consuming appliances, equipment and buildings.

Measures for the improvement of energy efficiency include financial instruments (subsidies, tax breaks on construction permits for heat insulation work on buildings, co-financing of renovation work) and mechanisms to encourage energy efficiency (energy efficiency funds). Significant energy savings are also to be achieved through the activities of ESCO-style energy companies. In this context, the plan advises the drawing-up of legislation to encourage the development of ESCO companies in Romania. As a member state of the EU, Romania also benefits from co-financing through structural funds.

Regarding renewable energy, since Romania has significant fossil fuel resources (coal, oil and natural gas), the development of renewables has not been a primary concern until now. Having plenty of renewable resources, Romania has currently only developed hydropower in significant amounts (currently about one third of the total electricity is supplied by hydropower). Under the draft EU Directive on RES, the target to be achieved in 2020 is 24 per cent of final energy consumption. The main potential for renewable energy production is from biomass, hydro, wind and solar sources. Government Decision (GD) 1892/2004 introduced a quota system for new RES for electricity production (Green Certificates). Feed-in tariffs for RES are relatively modest, but high for autonomous small wind systems (up to 110- 130 EUR/MWh).

Financing schemes

EU structural funds

The EC pre-accession aid to Romania increased significantly during 2004-2006 to over €2.8 billion for this three year period. Roman ia became a member of the EU on 1 January 2007 and, following on from pre-accession assistance, it continues to have access to the post-accession funding. Romania has prepared a series of strategic and operational documents to establish the development priorities of Romania and the allocation of EU funds over the next seven year period.

Romanian Energy Efficiency Fund

In 2001, the Ministry of Finance and the World Bank launched the GEF Energy Efficiency Financing Facility Project (FREE). The Fund started to work effectively in 2004. FREE aims at providing loans for commercially viable energy efficiency projects. GEF seed money ($ 10 million, of which $ 8 million for investment and $ 2 million for technical assistance) is orientated at leveraging co-financing by Romanian and foreign sources. FREE is designed to have a demonstrative effect and increase the banking sector's interest in investment in the field of energy efficiency in Romania.

In 2008, based on the successful results of energy efficiency and renewable energy credit lines in neighbouring countries, EBRD launched its first credit facility for the financing of energy efficiency projects by private industrial companies in Romania. This is part of the EU/EBRD Energy Efficiency Facility, a wider joint programme of the European Commission and the EBRD. Funds of up to €80 million will be provided through the programme to banks involved in lending for energy efficiency projects in Romania. Complemented by €24 million in EU grant funding from the Phare programme for technical assistance to support energy efficiency projects and provide incentives for new investments, the Facility will provide an integrated package of loans, grants and technical assistance for industrial companies implementing eligible energy efficiency measures.

In January 2008, EBRD claimed that three loans had already been signed: €20 million for the Banca Comerciala Romana, €10 million for CEC and €5 million for Banca Transilvania. In May 2008, the EBRD agreed a €10 million loan to Romania's BRD Groupe Société Générale for on-lending to private industrial companies in the country to finance energy efficiency projects.

ESCOs

ARCE organized many dissemination actions (seminars, publications in mass media) in order to present the importance and the role of third-party financing or ESCOs, in the field of energy efficiency investments.

In 2003, the Romanian American Investment Fund and EBRD capitalized an ESCO, the Romanian Industrial Energy Efficiency Fund, Romanian Industrial Energy Efficiency Company (RIEEC), a special purpose vehicle to be owned by the Romanian American Enterprise Fund (RAEF) and a Romanian energy services company, EnergyServ. In 2004 and 2005 RIEEC developed several projects, providing financing for energy efficiency investments in creditworthy industrial companies, for on-site co-generation systems.

Banking sector

In the World Bank Doing Business 2007 report, Romania was ranked as the top reformer in Europe and the second in the world for the period 2005-2006.

However, according to the EBRD-World Bank Business Environment and Enterprise Performance Survey (BEEPS), in 2005, new investments in the corporate sector are mostly financed on internal funds. The median debt to asset ratio stills low, suggesting that firms are not becoming more leveraged.

The Romanian banking sector is in continuous development, both legislatively and operationally. The significant number of amendments to the banking legislation, as well as the change in the corporate structure of the big players on the Romanian banking market, show clearly that the aim is the harmonization with the relevant European legislation and the implementation of the new concepts of corporate governance.

The starting point of the legislative changes brought to the Banking Law consisted of the explicit addition of new concepts, such as: credit institutions, financial institutions, electronic payments, issuers of electronic payment instruments, etc. The supervision of banks on a consolidated basis and in cooperation with the regulatory authorities from the countries of the parent companies, the changes in the regulations dealing with the prudential indicators and the calculation thereof, more detailed provisions on banking secrecy allowing the provision of information for the purpose of a more efficient supervision of banks are meant to create a framework consistent with the regulations applicable at the EU level. The amendments brought to the banking legislation had mainly in view the improvement of the means of the National Bank of Romania (NBR) with respect to prudential supervision of the banking sector, as well as the implementation of the provisions of certain European Directives.

At an operational level, it is worth mentioning the creation by a large number of banks of the Credit Bureau by putting together an information sharing mechanism with a view to identifying and assessing credit risk and protecting creditors and credit market by reducing fraud and at the same time of giving, to a certain extent, consumers the possibility of taking benefit of a positive credit history. Further, to the same end, NBR has created specific norms to regulate the granting of consumer credits.

In the context of the new legislative framework and the positive approach taken by the NBR towards cooperation with the corresponding authorities from the EU member states, the operation of banking activities in Romania seems to follow a clear increasing trend.

Despite these major advances in the financial sector in recent years, the intermediation of savings remains relatively low and bank lending still only accounts for 37.9 per cent of GDP (according to EBRD). The level of intermediation in rural areas appears to be especially low; according to the EBRD/World Bank Life in Transition Survey, less than 10 per cent of households in rural areas have a bank account (40 per cent in urban areas have one), compared to more than 50 per cent of rural households in Hungary and Poland and more than 70 per cent in the Czech Republic and Slovakia. Privatization in the banking sector is nearing completion. However, the privatization of CEC (Savings Bank) has been postponed after the government rejected the sole bid submitted by the National Bank of Greece, on the basis that the offered price was too low.

Brief conclusions and recommendations

- Within the harmonization process with EU legislation, EE is a matter of high priority at national level.
- Primary and secondary legislation for EE is developed. However, measures have recently been introduced and relatively little is yet known in terms of impacts.
- As Romania has significant fossil fuel resources (coal, oil and natural gas), the development of RES has not been a primary concern until now.
- At the company level, the credit available to the enterprise sector is mostly debt finance (nearly all bank loans, but also corporate bonds in selected cases).
- Financing schemes are mostly on-balance sheet financing, or sovereign guarantees for large state-owned projects.

Energy efficiency and renewable energy sources project development and finance capacities

Existing and prospective EE and RES projects

Several projects are active under EU structural funds, the Romanian energy efficiency fund and the Romanian energy efficiency and renewable energy credit line.

Also, two projects have been identified under the Joint Implementation (one of cogeneration and one of district heating system)

There is no opportunity to provide with the assessment of local capacity for project development and financial project analysis capacity building due to the fact that no assessment mission could be held to Romania.

However, based on the knowledge about existing capacity in Romania to identify and assess potential EE or RES projects, the relatively low activity in EE and RES is observed outside existing support schemes from international cooperation. A capacity to develop fully bankable project proposals corresponding to financial institutions' standards and practices is rather restricted to a limited number of actors.

Assessment of equity and mezzanine financing business development skills

The country is now viewed as holding some of the region's greatest potential for private equity and has attracted substantial interest from private equity funds. Many regional funds maintain an office in Bucharest (including Mideuropa Partners, Advent, Southeast Europe Fund, GED etc.).

However, regarding EE and RES, the experience from RAEF managing the Romanian Industrial Energy Efficiency Fund showed that project finance was still not very well understood in Romania. Financing schemes are mostly on-balance sheet financing, or sovereign guarantees for large state-owned projects. There is a lack of preoccupation and competence/experience in energy efficiency project finance at local banks level (assessment procedures, application forms, etc), which prefer to rely on credit line schemes, as shown by the rush for projects by intermediary banks after the EBRD scheme was launched.

Brief conclusions and recommendations

Existing projects show that some technical skills exist in Romania, both on energy efficiency and renewable energy issues. A number of different Romanian organizations have participated in EU funded projects, including regional and municipal energy and energy efficiency agencies, universities, NGOs, etc. Many of these projects have supported the capacity building of these organizations, and have helped to raise the awareness of the authorities and the general public.

The experience from RAEF managing the Romanian Industrial Energy Efficiency Fund showed a lack of awareness by management of industrial units, local authorities, on technological solutions available and energy efficiency project management principles. It also showed a lack of sufficient investment project management, energy performance contracting and engineering experience due to lack of projects in the clean energy sector.

Stimulated and supported by the EU accession process, the funding sources for energy efficiency activities in Romania should evolve and shift from donor programmes to Romania's own funding sources, for example from the state budget and, most importantly, the EU structural and cohesion funds. The schemes on public-private partnerships, like the FREE or EBRD Credit Lines, are a good signal towards a generalization of EE financing practices. A crucial issue would be to extend project realizations without such support schemes on a traditional business basis.

Investor interest

No mission could be held in Romania. The assessment is based on existing documentation. Therefore no potential investor interest could be assessed. However, existing activities and regulatory framework suggest that EE and RES are a matter of concern. Identified actors should be contacted during future project activities.

Energy overview

Russia is a global energy player with the largest natural gas reserves in the world and substantial proven oil reserves. In 2006 Russia provided more than 12 per cent of global crude oil production and about 22 per cent of natural gas production, becoming the world's second largest producer and exporter of crude oil and the world's largest producer and exporter of natural gas. Russia is also a leading global producer and exporter of petroleum products, coal and electricity.

Russia's strengths in the energy sector are massive gas and oil reserves, proximity to the European and Asian markets, well developed pipeline and sea terminals network serving European markets and substantial control over the export routes for the Caspian Sea region energy resources.

Following a period of rapid increase during 1999-2004, oil output has seen more modest growth in recent years (around 2 per cent annually) and this rate is expected to continue for the next few years.

Natural gas production has been increasing by 1-2 per cent annually for the last few years. Over 80 per cent of the total production comes from Gazprom, which also controls the pipeline network and has an export monopoly.

Russia controls a very well developed export pipeline infrastructure to supply the European market, although it needs maintenance and possibly expansion. It is actively pursuing a policy of limiting reliance on transit countries (mainly Ukraine and Belarus) in exports, by building new pipelines and sea terminals.

For the last few years Russia has seen a strengthening of governmental control over the oil and gas sector, limiting the role of foreign investors and actively supporting the state-owned companies Gazprom and Rosneft.

Difficult geographical and climatic conditions in the most promising producing regions, ageing, under-invested pipeline infrastructure and an unreformed gas sector are among the major weaknesses.

Opportunities relate to improving energy efficiency, curbing domestic energy demand and limiting losses, restructuring and tariff reforms in natural gas, electricity, and 20 transmission sectors to stimulate investment, entering new energy markets (particularly in Asia) and development of liquefied natural gas technology.

Risk factors are related to the consequences of fast increasing state ownership in the oil industry, a worsening FDI climate and deterioration of political relations with the EU.

Russia is characterized by high energy intensity. Its per capita consumption of primary energy is close to the OECD average, but national estimates put the energy saving potential of the country at 360-400 million toe, which is equivalent to 30-40 per cent of total Russian energy consumption.

The energy market structure varies in the regions – some are predominantly coal, oil, or gas oriented, some have several of these sources. The market structure in each of the regions is decided by the availability of energy resources and by the energy demand. Each of the regions has its own specific conditions, but usually large federal level energy companies are present in each of them. The regions are very different from each other and separate studies are needed for each of them.

The state policy on tariffs is implemented by the Federal Agency for Tariffs, which approves minimum and maximum tariff for the energy producers. The concrete tariffs in the oblasts (regions) of the country are decided by the regional energy commissions in accordance with the local conditions. The tariffs may be several times bigger or smaller in different regions.

The tariffs for electricity and heating are different in different regions. Social sphere users (hospitals, schools etc) benefit from preferential tariffs in comparison to other users (for example industry). The structure and amounts of preferences are the prerogative of the respective regional energy commission, which sets them in accordance to the local conditions.

Existing energy resources, energy dependence for primary and secondary energy resources, production of electricity and of heat, use of renewable energy sources (RES)

Russia's power transmission is organized as a unified power system that connects 70 localized energy systems and allows the transfer of power across the country. This unique situation is favourable to renewable energy projects in remote locations with access to transmission facilities, which can deliver power to more densely populated areas. The size of the Russian territory constitutes a strong potential for the development of renewable energy sources.

- Russia has excellent potential for wind power generation. An attempt to utilize just 25 per cent of its total potential would yield some 175,000 MW of power. The highest wind energy potential is concentrated along seacoasts, in the vast territories of the steppes and in the mountains.
- The overall technical potential of biomass is estimated at 35 million toe which, if converted to electrical power, could generate nearly 15,000 MW. This includes sewage sludge, cattle manure, and lumber waste. With the reconstruction of pulp and paper plants, the use of wood waste is also becoming more prevalent.
- Similarly, hydro potential is quite large as 9 per cent of the world's hydro resources are concentrated on Russian soil. Hydropower provides 21 per cent of total generating capacity and there is a large potential for small to medium-sized hydropower projects.
- Geothermal potential is also high, with theoretical resource estimates of high temperature (>90 C) steam, water and brine at about 3,000 MW.
- Solar potential is reasonable despite the country's location in the northern latitudes. The highest solar potential is in the southern regions.

This renewable potential should be treated with caution, particularly the wind potential. Given the well developed grid, climatic conditions and patterns of energy use, the utilisation of intermittent resource, like wind energy, which also requires long transmission lines may not be economically and technically justifiable and may not, in the short term, obtain the necessary governmental support to make it financeable.

The development of renewable energy projects also faces barriers such as the lack of a legislative mandate (renewable portfolio standard or feed-in prices), low electric and heat tariffs, low public demand, barriers to the grid access, and the overall lack of investment capital. Nevertheless, the magnitude of the renewable energy potential qualifies Russia as a leading candidate for development.

Level of priority given to EE and to RES in country's energy policy

The President of the Russian Federation declared energy efficiency as one of the priorities of the country - Decree of the President of the Russian Federation from 04.06.08 № 889 on measures to increase the energy and environmental efficiency of the Russian economy.

Investments in the energy sector

According to Rosstat, in 2007 Russia attracted $28 billion of FDI, twice the 2006 level. The main sphere of investments was the energy sector (almost half of the total). Foreign investors were attracted by the reform of the electricity sector (RAO UES) and sale of YUKOS assets. In the electricity sector, the two biggest deals amounted to $7.7 billion (E.On acquired a share in OGK-4, and Enel in OGK-5), and in the gas sector Enel and Eni bought Yukos assets for $5.8 billion.

Among the most important sources of inward FDI were Cyprus and the Netherlands (countries of domicile of many Russian investors) and also Germany, the United Kingdom and the United States. OECD experts found that Russian laws on FDI are softer than in China and India but significantly stricter than in OECD countries. They estimated FDI restrictions in 29 OECD countries and 14 non-OECD countries – *FDI Regulatory Restrictiveness Index*. The obstacles for FDI were evaluated on a scale from 0 (full openness) to 1 (prohibition of FDI). The score for Russia was 0.318, compared to 0.401 in India, 0.405 in China and a 0.144 OECD average.

Many national and foreign entrepreneurs notice an unfavourable investment climate in Russia (which is reflected in many international research works).Among the negative factors one can mention problems related to property rights, unpredictable "rules of the game", high corruption (121st place among 163 countries, according to Transparency International, 2007). High administrative barriers, bureaucracy, expensive and time consuming regime of permits, unreformed court system, lack of transparency are also of high importance.[8]

It has been estimated that at least 20,000 MW of new generating capacity will be needed in the next four years. In fact the country added only approximately 2,000 MW in the last 18 years.

At present (January 2009), all investments in the sector have practically ceased. As past performance provides little guidance to the future, it is difficult to predict, when and how the investments will resume and who will be the future investors, as only the multilateral financial institutions like EBRD, EIB and IFC can be considered as potential investors in the current credit crisis.

---

[8] Based on Rosstat data, OECD (2007), http://www.napi.ru/runews/runews_407_5.aspx; http://www.vedomosti.ru/newspaper/article.shtml?2008/01/14/139545.

Financial environment in energy efficiency and renewable energy

- The economy may be further affected by the spread of the global financial market crisis in late 2008. The state retains important reserves to help stabilize the situation, but macroeconomic stability will be threatened if oil prices keep falling while access to external financing is restricted. The possibility of sharp rouble devaluation is the greatest risk to macroeconomic stability in the short term.
- The budget surplus is forecast to shrink rapidly, in the light of lower forecasts for oil prices, slowing economic growth and substantial budgetary outlays in support of the struggling financial sector.
- The impact of the global financial market turmoil, a global slowdown and decreasing commodity prices may lead to a deceleration in economic growth in 2009. Growth should pick up thereafter, but will continue to be constrained by the slow pace of institutional change. The Economist Intelligence Unit forecasts that real GDP growth will average around 4.7 per cent per year in 2010-13.
- Despite problems in the business environment, ample market opportunities have led to large inflows of FDI. However, the growth of FDI inflows could slow in the first half of the forecasted period, and even by 2013 the stock of inward FDI could still be equal to just 19 per cent of GDP.

Legal, regulatory and policy framework

The main governmental institutions involved in the development and implementation of policies and measures in the fields of energy, energy efficiency and renewable energy are: the Federal Council (respective committees of the Federal Council and the State Duma (Parliament), the Government of Russia, and the Ministry of Energy. In the entities of the Russian Federation they are the respective governmental and executive bodies.

The main regulatory laws in the field of energy, energy efficiency and renewable energy are the federal laws and Governmental regulations:
- Federal Law on electrical energy dated 26 March 2003;
- Federal Law on functioning of electrical energy dated 26 March 2003;
- Federal Law on energy savings dated 3 April 1996.

Governmental regulations include the following:
- Regulations on provision of communal services dated 23 May 2006;
- Regulations on regulation of access dated 27 December 2004

There are no specific norms and standards on energy efficiency.

Currently the Ministry of Energy has developed a draft Concept for increasing efficiency of energy consumption in Russia until 2012, which is undergoing an approval procedure. After the approval of the Concept, a corresponding Target Federal Programme will be developed.

The main policy documents (strategies, programmes, action plans) in the areas of energy and energy efficiency are:
- Energy Strategy of Russia for the period until 2020 – approved by the Government of the Russian Federation on 28 August 2003;
- Decree of the President of the Russian Federation of 4 June 2008 on measures to increase the energy and environmental efficiency of the Russian economy;

There is no information on programmes related to energy efficiency developed for specific sectors of the economy. Recently a programme for the Russian Railways was adopted.

There are no separate laws regulating renewable energy. The development of RES is considered as one of the methods for increasing energy and environmental efficiency and is one of the directions of the energy strategy.

On 4 November 2007 the Federal Law № 250-ФЗ on amendments of legislative acts of the Russian Federation related to reformation of the united energy system of Russia was ratified. The Law contains a number of measures to support electricity generation from renewable energy sources. However, most of the incentives and bonuses are applicable only to the participants of the wholesale market for electricity. The generation from small sources is practically excluded from the incentives.

Interest in receiving equity and mezzanine financing

The interviews with the potential investors and financial institutions in Russia revealed that at present lending to new clients has practically ceased, but banks like EBRD, IFC and World Bank are in the process of preparing new financial facilities.

The provision of mezzanine debt is closely related to the availability of senior debt. Should the planned facilities of the international banks materialize next year, there will be demand for mezzanine debt.

Barriers to financing EE and RES

There are a number of issues affecting the interest, capacity, willingness and possibility to invest in renewable energy and energy efficiency projects in the current financial climate.

- In the current circumstances of the financial crisis, the efficient use of energy and raw materials will be even more important, but the lack of credit on the market and the falling prices of energy will make the investment in energy efficiency less attractive.
- In a situation of contracting economy and the need to reduce public spending, Governments will be less inclined to offer further subsidies for renewable energy, or to put in place schemes such as feed-in-tariffs for support of renewable energy, as they would lead to further increase of the tariffs for energy to the population and the industry. Furthermore, such large investments would require large public spending, or investment by private individuals.
- The legal and fiscal mechanisms for converting the savings into revenue that could be used to service loans are not yet fully developed, despite the existence of the respective framework legislation.
- The borrowing capacity of the municipalities is low and will get even lower in the short term. Collection of money from private individuals affected by the economic crisis for improvements of private dwellings would be problematic.
- Most of the rehabilitation projects are relatively small – from $100,000 to $1-2 million, except the major projects in big cities like Moscow, St. Peterburg, or Rostov. Some of the measures such as replacement of pipes for hot water would take a long time and require high investments, which would make measures hard to justify in the framework of a commercial project.
- Short term loans practised by national banks while EE and RE projects require medium and long term loans.
- The subsidized nature of domestic tariffs for electricity and heat – i.e. the tariffs are set by a government agency and can be set arbitrarily to meet current political objectives, and not in the interest of the investors.

- Difficulties in monetizing the realized savings – for example, savings realized in a budgetary enterprise mean that the money saved would not be disbursed to the enterprise and would remain in the Government's budget, but could not be used for repaying the investment.
- Investing in projects owned by State and municipal entities always raises the issue of guarantees and collateral – can the project be guaranteed by the State (slow and cumbersome process, not practical for small projects < $ 50-100 million), what is the value of such a guarantee if the credit rating of a particular country is low, can a municipal and public property be accepted as a collateral and many other similar legal and financial issues related to that.
- Significant obstacles and barriers to the grid access that a potential investor faces in the power generating sector.

Financing schemes

EBRD, in cooperation with the Russian Federation and its key Ministries, as well as donors, is in the process of establishing the Russian Sustainable Energy and Carbon Finance Facility (RSECF) that will channel financing through participating local commercial banks ("Participating Banks" or "PBs") competing with each other on loan terms and conditions. A limited number of PBs would then on-lend for sustainable industrial energy projects.

IFC project initiative is to promote investments in energy efficient technologies in Russia. IFC is to create an investment facility to provide credit lines for Russian financial institutions who will on-lend to energy efficiency and renewable energy projects. The credit lines will be supported by a partial risk guarantee, funded by GEF. The investment facility will operate in parallel with an extensive technical assistance programme. IFC will invest initially $ 20 million in the financing facility. GEF will invest $ 2 million in a guarantee facility to support portfolios of sub-loans. The cost of the technical assistance programme will be $ 6.25 million funded through GEF, donors and using IFC's own internal resources.

At present Russian banks actively use a syndicated lending as one of the main sources of financial provision. For example, the programme of the Russian Development Bank (RusDB) for financial support for the development of SMEs is based on a two-tier lending procedure, which is widely used by the development banks of other countries. RusDB provides financing to the partner banks, which on-lend money to their clients (SMEs), and to the organizations, developing infrastructure of the SME support. The participants in the programme are as follows: RusDB, regional banks, legislative and executive authorities, public organizations, subjects of small and medium-sized entrepreneurship. The volume of loans provided to SMEs by RusDB as of 1 July 2008, amounted nearly to 7.5 billion roubles. The loans are provided for a time period of up to 10 years. RusDB works with 90 partner banks and main requirements for these banks are as follows:
- Central Bank of Russia license
- Minimum 5 years of operation period
- Own capital of €5 million or more
- Positive audit report for the last accounting year
- Compliance with Bank of Russia statutory requirements
- Participation in deposit insurance system
- Absence of the Bank of Russia sanctions in form of prohibition on execution of some bank transactions and opening of affiliates/branches, as well as in form of suspension of license for some bank operations
- Size of own capital, calculated using Bank of Russia procedures, shall not be lower than the authorized capital stock
- No losses for 3 consecutive accounting periods

- Positive business reputation of the bank, its shareholders/participants
- Minimum 2 years of experience in crediting small and medium-sized business;
- Minimum loan request equal to 30 million roubles.

The following factors are also taken into consideration:
- Multi-branch banking
- Experience in crediting operations with small and medium-sized business
- Operational risk management system.

The loan size for a partner bank is determined based on its financial performance; it may reach up to 30 per cent of the working capital.

The largest part of the RusDB loan portfolio in 2007 (47 per cent) consisted of loans with a term of up to 3 years. The structure of the portfolio includes industry (36 per cent), trade (31 per cent), services (17 per cent), and construction (13 per cent) SMEs. Currently RusDB lends to partner banks at a rate close to the Central Bank refinancing rate, with the weighted average rate as of January 2009 being 13.03 per cent annual. Final on-lending rate is determined by the partner bank independently. The range of interest rates for small and medium-sized business varies from 11 per cent to 22 per cent, with the average annual interest rate about 16 per cent. During implementation of the programme there has never been the case where the limits allocated for regional bank funding have been used completely. This situation demonstrates that the partner banks are not able to find a sufficient number of sub-borrowers in order to use the limits set for them by RusDB. According to the 2008 data, the size of private sector in Russia amounts to 17 per cent of its GDP.[9] Shortage of the client base is explained by the relatively low activity of small and medium-sized business in Russia in seeking outside financing. The lessons learned from the RusDB operation should be considered in setting up and operating the Fund.

In the area of EE financing in Russia, one programme is implemented by IFC. Under this programme begun in 2006, IFC provides stand-by loans to the Russian financial institutions, which finance commercial investment projects in the field of EE and RES. In addition to these loans, IFC provides guarantees to the banks in order to promote funding for such types of projects. This enables IFC partner banks to issue sub- loans under general market terms (about 10 per cent annual in currency). The projects payback period is up to 7.5 years, maximum size of one loan is $2 million. The borrower has a chance to be granted a one year grace period in payment of the main debt amount.

The issues above lead to a preliminary conclusion that an equity investment in state and municipal owned entities is practically impossible, but this needs to be investigated further and be affirmed from a legal point of view.

ESCOs

There are number of ESCOs operating under different external supporting programmes (EU, Scandinavians etc). Some of the best known are: Center for Energy Efficiency, Center for Energy Policy, AcademEnergoServis, Institute for Energy Policy, RusDem, Regional Centres for Energy Efficiency in Kaliningrad, Murmansk, Kola, Karelia, Ekaterinburg etc., ESCO Negawatt, Rus Esco, 3E, Energo Servis.

The mission however, does not have reliable information on their financial performance and results. An IFC study (Energy Efficiency in Russia, 2008) has concluded that ESCOs have had "limited success in Europe" and suggests that the use of energy performance contracts be considered.

---

[9] IA "RosFinKom", http://rosfincom.ru/news/18694.html.

The Ministry of Energy informed the mission that the establishment of a state ESCO "Federal Service Company" had been planned. It is envisaged that it will create a network of ESCOs in the regions to cover all Russia territory with their activities

An equity investment in a particular project, where a special purpose vehicle is established and the State is one of the shareholders, may be possible, but needs to be confirmed. However, this approach is not cheap and may not be very practical for smaller projects.

To overcome this problem, some companies employ quasi legal schemes, where an informal partnership is established and the profit from the project is split between the partners in a sort of "gentlemen's agreement". However, such a scheme is built on trust and it is doubtful that a foreign entity, like the Fund, would agree to such an informal way of doing business.

Investments in ESCOs are possible, but it has to be noted that an establishment of an ESCO requires a lot of time and usually soft funding. For example, EBRD has spent almost 10 million dollars of soft funding (mostly grants) to establish UkrESCo in the Ukraine.

The financial results of the operation of ESCOs in Eastern Europe are not very encouraging, but they have established a presence and accumulated considerable experience and contacts. One of the options to invest in smaller projects and ESCO is to buy equity in one of the established ESCO. Financing an establishment of an ESCO from the initial stage is not an option for an investment fund, which does not have grant funds.

Banking sector

The Russian banking sector has about 1000 operating banks, mainly concentrated in Moscow and St. Petersburg (84 per cent of total assets). Scales of the Bank sector have dramatically increased over recent years. In 2007 bank system assets have increased by 44.1 per cent (in 2006. — by 41.1 per cent, in 2005 — by 36.6 per cent), reaching 20,125 billion roubles. Only in 2007 assets of credit organizations ratio to GDP have increased by 9 percentage points and reached 61 per cent.

At the same time, the Russian banking sector still lags behind leading world economies in absolute indexes. In Russia, banks play a leading role in the financial system, although in developed countries the index of bank assets size ratio to GDP is considerably higher: about 300 per cent in Germany, 250 per cent in France, 360 per cent in the United Kingdom.

The Central Bank of Russia refinancing rate is 13 per cent and the loans for capital expenditure are issued by selected banks.

The loans for fixed asset acquisition have interest rates starting from 10 per cent in euros, 10 per cent in dollars, and starting from 12 per cent in roubles. The prevalent rates are over 14 per cent with loan term less than 5 years. Normally, regional banks have higher interest rates than Moscow credit institutions for such types of loans.

Brief conclusions and recommendations

The macroeconomic situation in Russia is currently dominated by the global credit crunch. Industrial output, generally an early indicator of GDP trends, has been falling over recent months. And the output decline appears to have accelerated dramatically in November and December 2008. According to the latest government figures, manufacturing output were reduced by an additional 10 per cent by the end of 2008. (The situation with Russia's service industries, which is not recorded in the industrial output figures, is even worse).

In many sectors, the Russian companies are reporting declining orders and lay-offs. Construction, banking, metallurgy and the automotive industry are all in deepening crisis. Russian railways have reported a 20 per cent decline in freight volumes, reflecting the nationwide reduction of industrial production. The State provided emergency loans to a number of large companies. Some economists and businessmen predict a contraction of Russia's GDP by 5 per cent - 10 per cent in 2009. Some even talk about "severe depression".

The interviews with the potential investors and financial institutions reveal that at present lending to new clients has practically ceased, but banks such as EBRD, IFC and World Bank are in the process of preparing of new financial facilities (see above).

The provision of mezzanine debt is closely related to the availability of senior debt. Should the planned facilities of the international banks materialize in 2009, there will be demand for mezzanine debt.

Energy efficiency and renewable energy sources project development and finance capacities

Existing and prospective EE and RES projects

Projects focused on energy efficiency

GEF / UNDP:
- Capacity building to reduce key barriers to energy efficiency in Russian residential buildings and heat supply €2 900 000, 1998-2006;
- Reliable, energy-efficient municipal utility services (focuses on modernization of the current system of relationships in covering municipal utility costs in the city of Cherepovets) €100,000, 2002-2006;
- Cost-effective energy efficiency measures in the Russian educational sector € 1,700,000, 2002-2006;
- Standards and labels for promoting energy efficiency in Russia €16,000,000, 2007-present;
- Financing energy efficiency in the Russian Federation €19 000 000, 2003-present;
- Energy-efficiency investment project €1,250,000, 2000-2005.

GEF / IBRD / IFC:
- Promotion of energy efficiency and greenhouse gases' emissions reduction in combustion plants n/a 2004-2005;
- Greenhouse gas reduction project €2 200000, 1995 -1999;
- Gas distribution rehabilitation and energy-efficiency project € 66 250 000, 1995-2003;
- Municipal heating project for the Russian Federation €80 000 000, 2001-2008;
- Russia: Housing and communal services project €13 0 000 000, 2008 -present.

UNECE:
- Energy-efficiency investment project development for climate change mitigation (Belarus, Bulgaria, Kazakhstan, Russian Federation and Ukraine) €1 250 000, 2000-2005;
- Financing energy efficiency and renewable energy investments for climate change mitigation (Albania, Belarus, Bosnia and Herzegovina, Bulgaria, Croatia, Kazakhstan, Republic of Moldova, Romania, Russian Federation, Serbia, the former Yugoslav Republic of Macedonia, Ukraine) €6 900 000, 2007-p resent.

Global Opportunities Fund (GOF):

- Promotion of energy efficiency and GHG emissions reduction in the glass manufacturing sector n/a 2005-2006;
- Role of improved Russian energy efficiency in economic growth and energy security n/a;
- Reform of the Russian district heating sector n/a;
- Development of the sustainable energy and investment training programme for the power generation and power consuming sector n/a until present;

EuropeAid Tacis Projects:
- Steel sector energy efficiency programme in the Russian Federation €474 000, n/a;
- Energy efficiency at the regional level - Arkhangelsk, Astrakhan, and Kaliningrad Regions €3 000 000, 2006-2007.

EBRD:
- Sakha (Yakutia) Republican MSDP GUP ZhKH, €71 000 000, 2005-present;
- Ufa district heating loan €15 000 000, 2005-prese nt;
- Kaliningrad water and environmental rehabilitation: district heating € 11 200 000, 2007-present;
- Khanty-Mansi municipal services development programme €1000000, 2007-present;
- Irkutsk regional heating programme €28 600 000, 2 007-present;
- Lipetsk municipal infrastructure project €20 000 000, 2008-present;
- Moscow district heating loan €200 000 000, 2008-p resent.

USAID (ROLL) Programme for Sustainable Development of Model Communities:
- Energy efficiency project supports sustainable community development n/a 2005

REEEP projects:
- Financing municipal energy efficiency in the Commonwealth of Independent States € 200 000, 2006-2007;
- Barrier removal for residential energy efficiency €117 000, 2005-2006;
- Building energy-efficiency codes in Russia and Kazakhstan €230 000, 2006-2007.

Brief conclusions and recommendations

Technical engineering skills in the Russian Federation are very high and no additional training of technical skills would be needed. There is also sufficient experience in financial project development and preparation in selected companies and organizations in the country. At the same time, based on working experience in remote areas and cities in the Russian Federation, it may be concluded that additional awareness raising and capacity building for preparing bankable project proposals would be of value.

Hence, two types of capacity building activities have been identified as potentially beneficial to the selected target audience:

- Awareness raising seminars for decision-makers that are not yet actively involved in the support of energy efficiency and renewable energy activities. An overview of business principles related to the development of bankable energy efficiency projects could be provided so that decision-makers could identify potential projects to be developed within their region or community and, when necessary, support the project developers with relevant regulatory reforms and delivery of needed permits or concessions.
- Dedicated training courses for project developers that are focused on preparation of business plans ready for submission to financial institutions. Such training would consist of 2 or 3 sessions and require personal work in-between the sessions.

The table below reflects various groups of target audience and specific capacity building that could be beneficial to them.

Target audience	Adapted capacity building
Large corporations and most of the local financial institutions have acquired significant capacity to develop and present business plans on an international level. Large energy and industrial companies that are operating on the international market, have issued bonds and completed initial public offerings and their top management is perfectly capable of procuring the preparation of a business plan in accordance with the international standards. A complete business development programme is not needed, however management of such companies would usually welcome support in structuring projects in line with particular criteria or requirements of a specific financial instrument, such as a mezzanine debt provider.	**A session on mezzanine debt and specialized instruments** Assistance could be provided on incorporation of details related to specific mezzanine debt requirements of the Fund into the business plans.
Smaller industrial and municipal companies or less experienced project developers may have some knowledge and understanding of preparation of a business plan in a form of Technical Economic Feasibility, but usually this is not sufficient to present the project as a bankable proposal ready for appraisal by a financial institution.	**General business plan preparation programme** Such organizations will need more detailed training and assistance in preparation and presentation of their business plans, with particular emphasis on documentation requirements, financial projections and modelling.
Some small, mostly individual developers and remote municipalities have very general understanding of the business cycle.	**Basic training** Training on the basics of business plan preparation may help identify opportunities and further develop projects.
Selected audience of local and national level managers and decision makers, including representatives from NC, NPI, etc. who are expected to facilitate further development of secondary laws, regulations, standards and related activities to overcome remaining barriers and accelerate the identification, selection and development of energy efficiency and renewable energy investment projects.	**Training for decision-makers** Seminars on financial engineering and business planning to provide clear and concise information on what are the general requirements of international financial institutions for an investment project. This would provide understanding and support on the managerial and decision-making level on the identification, selection and development of energy efficiency and renewable energy investment projects.

Investor interest

Public sector

There is a clear interest from the public sector in attracting investments to increasing energy efficiency. This position was confirmed by the Ministry of Energy and the Ministry of Economic Development during earlier meetings. However, direct investments from the state to the Fund are difficult to obtain as it would require a very long and complicated legislative process.

One of the potential investors is the Russian Bank of Development (RosBD) created by the Russian Government in 2000. It is a 100 per cent state owned bank that provides credits for the real sector of economy and prepares investment projects of state significance. In 2006, the bank's assets reached $1 billion and net profit increased by 45 per cent from 2005. The energy sector represented 10 per cent of the total financing.

In December 2006, the Russian Government decided to merge the Russian Bank of Development with VneshEkonomBank (Bank for Development and Foreign Economic Affairs) and RosEximBank (Eximbank of Russia) into a Russian Development Corporation. All the three banks are state owned. The Corporation will be responsible directly to the Government, have an authorized capital of about $3 billion (70 billion RUR) and will not need the Central Bank's license.

The Corporation will be responsible for medium and long-term financing of investment projects in the priority sectors defined by the Russian Government. Energy efficiency is one of these priority sectors. The Corporation will have a significant capital to invest in the development of the priority sectors of the Russian economy. A direct participation in the Fund however will be difficult for a number of institutional and legal reasons. The Corporation is potentially interested in co-financing projects with the Fund.

Private sector

Private banks' participation in the Fund is probably of little interest to them, since this would reduce their T1 capital. Given the investment practices in Russia, in order for a private bank to invest in such fund, the bank would insist on owning a majority stake.

Potential partners for project co-financing

A viable avenue to be explored could be to co-finance with the existing, or to be established, Russian investment programmes and funds, like the National Projects, Target National Programmes etc. For example, see above information on the Russian Development Corporation.

Feasible partners of the Fund in Russia are the IFI's (IFC, EBRD, EIB) and other international banks and Export Credit Agencies financing projects in Russia.

Brief conclusions and recommendations

One of the possibilities to invest, particularly in smaller projects, is to co-finance in conjunction with the credit lines and facilities of EBRD, EIB and IFC established in the countries of operation. One example is EBRD, which in cooperation with the Russian Federation and its key Ministries, as well as donors, is in the process of establishing the Russian Sustainable Energy and Carbon Finance Facility (RSECF) that will channel financing through participating local commercial banks ("Participating Banks" or "PBs") competing with each other on loan terms and conditions. A limited number of PBs would then on-lend for sustainable industrial energy projects. Such co-financing would save time and money on project development and project appraisal, as well as benefit from the pipelines and the political clout of these large financial institutions.

A relatively large potential for renewables has been identified. This should be treated with caution – in a situation of financial crisis and financial austerity, spending of large amounts of taxpayers' money on subsidizing tariffs for renewables may not be something which the government can afford. One of the other issues identified is a lack of adequate infrastructure (transmission lines, substations) to transport the generated energy to the customers, which also requires large investments.

SERBIA

Energy overview

Serbia produces 20 per cent of its oil and 10 per cent of its gas consumption, importing the rest mainly from the Russian Federation. Oil is refined domestically with two existing refineries. All consumed electricity is domestically produced, for two thirds from lignite and for one third from big hydro plants. Electricity prices are among the lowest in Europe (4.6 euro cents per kWh) and do not provide the necessary resources for investments. Significant losses occur in production and in distribution whereas the efficiency ratio in thermal production is not high. In the heating sector, district heating also observes important energy losses coming from outdated equipment while in many households electricity is used to produce heating. Among renewable energy sources, small hydro, biomass and solar are considered to have stronger potential than other sources, such as wind and geothermal.

Legislation is in place or under amendment for improvement, and engineering capacities are of a good level. However, the country's economy is still in transition process, and energy efficiency is not regarded as a highest priority. In the field of environment, waste and water management are regarded of a higher priority than EE.

All public utilities for production of coal and electricity are state-owned, whereas district heating companies are in the ownership of local communities. The two largest companies in the Serbian energy sector, EPS (Electric Power Industry of Serbia) and NIS (Petroleum Industry of Serbia) have been undergoing a privatization and restructuring process. In line with the agreed schedule, NIS was privatized at the end of 2008, Gazprom becoming its major owner.

For the past 18 years, very little investment has been made in the energy sector, which strongly needs modernization. Energy efficiency financing is rare, financial engineering skills are lacking at all levels and special purpose vehicles do not exist as yet while project pipelines have been established neither by municipalities nor by SMEs.

Financial environment in energy efficiency and renewable energy

Legal, regulatory and policy framework

The Energy Sector Development Strategy of the Republic of Serbia by 2015 was adopted in 2006 and identifies as major priorities continuous technological modernization of the existing energy facilities, increase of energy efficiency in production, distribution and utilization of energy, increasing use of RES, urgent investments in new power sources with new gas technologies and construction of new energy infrastructure facilities within the frameworks of amendments to the actual Energy Law, under the responsibility of the Ministry of Mining and Energy.

The Implementation Programme for the period 2007-2012 of this strategy was adopted in January 2007 and contains a separate chapter on EE and RE. The Programme identifies barriers to increasing efficiency in energy consumption and to widely using RE, recommending regulatory, policy, institutional, organizational and technical measures to overcome these barriers. It foresees the development of national regulations to establish favorable conditions for ESCO operation and the introduction of an energy passport system in buildings. The Programme is now under implementation.

Co-generation energy facility in the Clinical Center of Serbia

The Energy Law was adopted in 2004 and regulates the generation, transmission, distribution and supply of electricity, the organization and functioning of the electricity market, the transportation, distribution, storage, trade and supply of petroleum products and gas, and the production and distribution of heat, as well as establishing the Regulatory Energy Agency. Amendments and supplements to the Energy Law are currently under approval.

A law on the rational use of energy is under development and will introduce compulsory energy audits.

There is no existing energy efficiency law, however substantive EE provisions are set by the Energy Law and its amendments will bring additional provisions, such as the establishment of an energy efficiency fund (according to the Implementation Programme, Chapter 14), which is regarded as a necessary tool for increasing EE and stimulating rational energy use. The legal framework for the Fund's establishment should be created and respective secondary legislation scheduled for adoption in 2009.

Three decrees stimulate the use of RES in electricity production: decree on criteria and conditions for obtaining the status of privileged power producer and criteria for assessing their fulfilment; decree on a market model for the electricity privileged power producers; decree on measures for promoting electricity production by privileged power producers.

The Law on ratifying the Treaty establishing the Energy Community between the European Community and Albania, Bulgaria, Bosnia and Herzegovina, Croatia, Montenegro, Romania, Serbia, the former Yugoslav Republic of Macedonia, and the United Nations Interim Administration Mission in Kosovo pursuant to United Nations Security Council Resolution 1244, was adopted by the Serbian Parliament in 2006.

The Law on the ratification of the Kyoto Protocol was adopted by the Serbian Parliament in September 2007.

The Law on Integrated Environmental Pollution Prevention and Control entered into force in December 2004 and brings a new dimension for the industrial polluter to think about EE and RE use. The year 2015 is set as the deadline for existing companies to comply with this Law.

Interest in receiving equity and mezzanine financing

In the Ministry of Mining and Energy, those responsible for EE and RE showed strong motivation for the Fund, seen as an interesting opportunity to implement some projects, especially in the RE sector.

The private companies showed a good interest in receiving equity and mezzanine investments in EE and RE projects. A large number of RE and EE project ideas show a strong interest in using the Fund.

For the electricity utility, the Fund was seen as part of a solution to deal with the lack of investment during the last 10 years and a valuable tool for the installation of small hydropower plants that could cover the actual production shortage of 2 per cent that the utility is facing.

The main governmental petroleum and gas companies in Serbia are in a transition process due to privatization. They have big investment needs since no major investments have been made for about 30 years and they should meet the commitment of the Law on Integrated Environmental Pollution Prevention and Control for emissions reduction by 2015. No clear scenario will be known until the finalization of the privatization process.

The governmental Serbian Energy Efficiency Agency (SEEA) and the five regional energy efficiency centres are very interested in the Fund since they have been involved closely in many activities related to EE and RE. For these institutions, the lack of investment is a major barrier for project development and implementation especially for small hydropower plants where they can play a valuable role for Fund promotion.

Barriers to financing EE and RES

Energy saving potential is significant at all levels. However, project development faces barriers such as lack of information and awareness on the EE and RE financial benefits, lack of human potential such as lack of energy managers and lack of skills for project development or absence of obligation to conduct energy audits. Regulated electricity and district heat prices are considered to be below the market average (electricity – below 5 cents per kWh), while loans at an average interest rate of 12-13 per cent are considered as expensive. Available financial resources are very limited and most existing projects are of a rather small size while project bundling has not been practised yet.

In the public sector, a few institutions familiar with project development, such as the regional energy efficiency centres, can hardly spread their skills due to lack of resources for capacity building. Many municipalities have very little working staff and no staff responsible for energy. A lack of an intermediary structure between the national government and the municipalities (no regional level) leads to a lack of entity that could supervise energy issues for municipalities and to work on bundling EE projects between municipalities. Attention should be paid to the legal barriers preventing bundling of projects in the public sector where the ESCOs are expected to play a major role for project development for the Fund.

In the private sector, there are no incentives and programmes related to EE and RE projects. In many cases, EE is still not considered as a priority by private companies, who first want to meet international standards for exports and see investments in production as more beneficial while not always considering benefits they could receive from investing in EE improvement. Electricity prices for industries are even below the national average, which brings them lower than 5 cents per kWh. This missing leverage does not help to improve the country's energy intensity, economic development and private sector competitiveness. Since many companies are in the process of privatization, the needs for investment for infrastructure, clean energy production and energy efficiency improvement are being highly requested. The new EE fund and the feed in tariff that is planned to be in place in 2009 will shift the projects proposals into real project opportunities for implementation. The good potential for RE projects and the country's high energy intensity will put the Equity Investment Fund among the requested financing mechanism in Serbia.

The ESCO concept can be very helpful in bundling small size projects also in the private sector, however ESCO business can be risky since private companies are not very reliable in delivering and repaying benefits while the judiciary system is very slow to issue penalties, which results in lack of guarantees for the potential ESCO. The ESCO business has not yet received a strong interest from private companies because most of them offer very specialized equipment, are unable to offer the full modernization package and do not cooperate among themselves.

Incentives

Domestic resources of lignite are available for the next 50 years, however electricity supply capacity is at its maximum and in this respect the Government regards EE as a valuable resource for economic development. However, the country does not yet have clear economic or fiscal incentives for EE and RE promotion while the market is in need of EE and RE investments. The Ministry of Mining and Energy has analysed incentives used in the EU and feed-in tariffs are under approval. Several financing schemes have been established to provide grants and soft loans, but more incentives will need to be elaborated to facilitate commercial investments in EE and in RES.

Financing schemes

A few grants and soft loans are available and a few more are being put in place. No equity financing of EE and RES projects has been experienced until now. Financing needs are much higher than available resources.
- Fund for Environmental Protection and Energy Efficiency: Deriving its resources from pollution taxes and a little government budget input, the fund started operations in 2005. It has disbursed grants until now, but will be introducing revolving and soft loans schemes. No EE projects have been financed until now since priority is seen in purely environmental issues[10], but better cooperation could be foreseen. The fund seems to face capacity barriers: according to UNDP, only €3 million have been disbursed out of €15 million. The fund did not have any requests for financing EE and RE projects; however the Fund manager is keen to receiving similar requests.
- NIP – National Investment Programme: Established to finance mainly public sector projects of national interest, this fund is now managed by a special ministry created for it. Resources come from on-going privatizations and from the State budget. Financing is provided usually as a grant, sometimes also on a revolving basis. The priority sector for the next few years will be infrastructure. Energy efficiency projects submitted by the Ministry of Mining and Energy and by SEEA have been until now refused. The NIP allocated a mere 1.2 per cent of total funds to environmental protection in 2006–2007.
- Energy Efficiency Fund (KfW): This fund was set up by the German Ministry of Environment and the German Ministry of Economic Cooperation and Development

---

[10] For instance, Belgrade and many other municipalities are not equipped with any water waste management system, which threatens the condition of rivers and raises transboundary issues.

and was launched in the second half of 2008 through loans to four Serbian banks with a volume of €45 million and technical assistance for banking staff and customers,

- Energy Efficiency Fund (Government) – to be created: The working modalities of the Fund have been elaborated by SEEA and included in amendments to the Energy Efficiency Law. The Fund is expected to become operational after adoption of the amendments, tentatively by the end of 2009. The fund would be operational for 10 years and should each year receive about €20 million from taxes on energy consumption, state budget, international institutions and interest rates. About 20 per cent will be disbursed as grants, 50 per cent as zero interest rate or soft loans and 30 per cent in co-financing. Disbursements will be made into both THE public and private sectors on activities such as studies, training courses, awareness activities and projects.

- In 2007, EBRD signed a project financing up to €25 million for the private equity fund dedicated to investing in EE and RE projects in Central and South-Eastern Europe. The Fund invests in Central Europe, the Baltics, and South-Eastern Europe including Bulgaria, Serbia, Croatia, the former Yugoslav Republic of Macedonia and Ukraine. The Fund invests in projects that qualify as power projects that contribute to specific EU/Kyoto targets, and have predictable and legally binding long-term off-take agreements. Projects have known technologies and business models, exhibit scale large enough to be attractive for trade sale, and are managed by experienced developers, engineers and project managers. The Fund may also target strategic and proprietary infrastructure technologies and operations related to renewable energy. Regarding the structure, the Fund invests and co-invests in projects where equity returns can be enhanced by optimal degrees of debt using various financing structures. The Fund generally seeks a controlling position.

- Special financing schemes for EE, such as a project of the Raiffeisen bank, which offered a lower interest rate for a specific brand of isolating windows (2-3 per cent instead of the market 7-8 per cent).

- SEEA demonstration projects have been financed by donations, such as district heating modernization (€ 25 million, GTZ donation) or schools and hospitals modernization.

- Law on Investment Funds: further information will be provided by the Ministry of Finance.

ESCOs

Energy service companies do not exist in Serbia, mainly because there is no support mechanism in place for the work of these companies. The legal framework for running such companies is also lacking, but work has been done by GTZ to determine legal obstacles to the working of such companies in Serbia. The lack of governmental financial support (absence of an energy efficiency fund) makes the work of such companies practically impossible. Legislation and some capacities are under development and further legal and capacity advancement is necessary in order to fully enable third party financing.

During the discussion on the necessity of an ESCO for bundling small scale EE projects the coordination of ESCO creation through the Chamber of Commerce was proposed. The assessment of the previous situation (in the absences of ESCOs) and market needs for ESCOs and a support for this project through the NC and the NPI would be appreciated.

Serbia currently has 36 working banks in the country. The sector is competitive and carefully monitored and regulated by the National Bank. Serbia has a medium debt level, which is for instance much lower than in the neighbouring Croatia. The market is still regarded as emerging, but very open to western ways of functioning and, although it has a high risk ranking, real risk is much lower. EE projects are regarded as less risky than the average.

Société Générale would be sensitive to EE and RE projects and would be eager to discuss the topic further with ECE.

Brief conclusions and recommendations

- The financial environment for EE and RE in Serbia could be considered as promising considering the expected changes in the legal framework and feed in tariff establishment for RE. The number of potential projects in RE and the projects developments activities recorded in the country, suggest a great need for financing mechanisms flexible enough to fit the different developer needs and capacities. The Equity Investment Fund could be quite solicited since many new companies active in the field of EE and RE will not have sufficient capacity for projects financing and implementation.
- The Standing Conference of Towns and Municipalities, an organization representing 167 municipalities, could be a good partner for Fund promotion among the municipal sector. It presents a good partnership for project development, Fund promotion, event organization and training participation.
- Existing co-financing schemes should be approached as potential co-financing partners of the Investment Fund.
- The ESCO creation project with the involvement of the Chamber of Commerce should be followed up.

Energy efficiency and renewable energy sources project development and finance capacities

Existing and prospective EE and RES projects

Serbia presents a high number of RE projects identified by project promoters, and the private sector's high level of interest for project implementation reflects the potential level despite the lack of regulations and incentives. The EE fund planned to be in place by the end of 2009 with the feed-in tariff will certainly bring leverage to EE and RE activities in the country. The newly formed DNA and the possibilities of applying for CDM projects is seen as positive sign by the private sector to invest in RE projects, where biomass, wind, thermal and small hydro projects present good opportunities. Until now there have been no CDM projects conducted in Serbia. Currently the Ministry of Environment and Spatial Planning is preparing the Initial national communication to the UNFCCC and, with Ministry of Mining and Energy, a draft strategy for the use of CDM in Serbia. Consideration of the Equity Investment Fund for investment in RE projects will certainly bring an interesting addition to project realization.

TEAM BRUNSWICK GROUP, MHE Power Balkan, Head commerce, Kirka Boiler production and GBC ESCO have expressed great interest in the Fund for projects realization and investment possibilities. On the request of the consultant, they have submitted 16 projects proposals which show different levels of knowledge and capacities.

Another programme is the improvement of energy efficiency in public buildings through the Serbia Energy Efficiency Project, financed by the World Bank credit and loan. Nine World Bank project portfolios active in Serbia as of June 2007 and the above-mentioned EE project targeting the district heating and central government administration that aim to improve energy efficiency in heating buildings in order to make it affordable and enhance the functional and health environment of the users.

There is another programme for Improvement of energy efficiency in five district heating companies through modernization financed by a grant of the European Agency for Reconstruction (about €20 million), while KfW Development Bank is financing the improvement of EE in six district heating companies through modernization for about €27 million. This Project will be extended to six more municipalities during 2009 (cost €30 million). The switch to consumption-based billing is the most important step of institutional reform which is to be achieved though this project. According to the activities documents, KfW is planning for the creation of a regional fund for EE and RE for Serbia, Montenegro, Bosnia and Herzegovina, the former Yugoslav Republic of Macedonia and the United Nations Interim Administration Mission in Kosovo.

GTZ has been working with the Standing Conference of Towns and Municipalities since 2002 for the modernization of the municipal services mainly for waste management, water management and local energy efficiency. The co-financing provided by GTZ is very low and for very small scale projects. Since the programme ended in 2008, GTZ is looking for opportunities and conditions to launch a new energy efficiency project in Serbia. Besides the Standing Conference, GTZ is investigating on opportunities for programmes with the Ministry of Mining and Energy, Energy Efficiency Agency and other institutions relevant for the energy efficiency field. These programmes would aim at improving the legal framework, as well as enhancing institutional capacity at central and local levels, enabling them to apply new legislation efficiently. The project is also oriented towards enabling local self-government units for a strategic approach to energy policy as well as for the preparation of EE related projects that can be submitted to various domestic and foreign financial institutions. The project is expected to run for three years.

Assessment of investment project development skills

Public sector

At the ministries level, the resources are more concerned with administration than by project development. Based on the interaction with the resources, it seems that they do not have deep skills for project identification and development related to the EE and RE projects. They are very little involved in such activities and they hope for a capacity reinforcement at least for the basics for investment project development.

The Ministry of Mining and Energy would be interested in being informed on how to prepare a business plan and how to review a business plan, which will help them to evaluate project potential in the country. They would appreciate a focus on development of wind projects, which constitute a new area in Serbia. Additional information on feed-in tariffs that are under development would be welcome.

The participants from the Ministry of Energy and Mining indicated that besides the lack of awareness on the different levels and the requirement of a complete legal framework, they are experiencing a shortage in the human resources level (three resources for EE and one for RE). The same situation is indicated in the Ministry of Economy, Environment and in the Serbian Energy Efficiency Agency. They would welcome capacity since they all have to deal with energy issues through their fields of competence, however they do not have skilled personnel to recognize the benefits of EE and to evaluate bankable project proposals.

Serbian Energy Efficiency Agency (SEEA): The resources within SEEA have good experience in technical issues and energy auditing. The agency has skilled people that can perform audits since they have already carried out energy audits in governmental buildings and municipalities, however more in-depth financial engineering capacities are needed. The bankable projects include technical solutions, risk assessment, financial analysis and energy management including measurement and verification plans. SEEA might need capacity building for bankable projects without the technical training. It was indicated that a lack of ESCOs is a barrier for the EE and RE market evolution, and the capacity building for ESCO implementation is highly recommended.

Regional centres for energy efficiency: The resources within the five regional centres are high qualified people; most of them are professors with great technical capacities. However, they have expressed a need for project development and financial project analysis capacity building with focus on RE projects. The participants need more skills for measurement and verification procedures related to EE and RES projects as well as assistance for projects presentation to financial institutions.

The municipal sector through the projects completion within the Standing Conference of Towns and Municipalities has an appreciable knowledge about the energy conservation measures that could be implemented particularly in the biggest municipalities. On the other hand, the small municipalities still suffer from a lack of skills and resources at all levels, bearing in mind that more than 100 of the 167 municipalities have fewer than 40,000 inhabitants. The municipal sector needs more awareness and capacity building for EE and RE directed mainly at the administration level for project development. The procedure's review for bankable project presentation suggests training is needed for detailed investment grade audit realization, measurement and verification plans and project financial analysis. The other important issue that has been raised is the missing skills for project bundling and presentation that might need to be addressed as a priority for capacity building for the small municipal projects considering that bundling is a must for applying for the Fund.

Private sector

The private sector has strong technical skills for project implementation, although for project development including all financing aspects is the missing link for project realization. All the private companies questioned have good project ideas for EE and RE projects development, but none of the projects have been presented to financial institutions. The main reasons were the current unfavourable market condition for project financing and low profitability for RE projects without feed-in tariff or incentives. However, focusing more on the resources ability, the project promoter recognizes that they do not have all the skills to develop bankable projects, especially for extensive financial analysis and technical risk analysis mitigation. The ESCO concept has pointed out the missing drive in the market and the need for skills to develop projects and acquiring all the necessary abilities for good projects analysis and presentation on technical level, but also on the risk, financial and contracting aspects too.

The commercial financial institutions are not involved in any activities related to EE and RE projects evaluation since no concerns about this sector have been demonstrated. The commercial banks are not considering the EE and RE projects as a possible new business line that can be offered to their clients, and they did not have any requests for financing them. The problem could occur from the demand side where the clients do not present specific projects and from the bank side which is treating all projects with the same mould. There is a persisting misunderstanding on both sides that needs to be addressed and solved by increasing the awareness level.

From the private sector point of view, the commercial financial institutions do not offer attractive loan conditions and apply the same rules to all project types. They do not take the contract into consideration as guarantee or collateral. The development of a guarantee fund might help the financial institutions to have better business understanding and create a more dynamic market for EE and RE project implementation. Commercial financial institutions would welcome additional capacity building related to specific evaluation of energy efficiency and renewable energy projects.

Assessment of equity and mezzanine financing business development skills

Very few people are familiar with equity and mezzanine financing and all of them belong to finance and investment institutions. Beyond the concept, they affirm having all necessary resources for project evaluation based on the equity and mezzanine scheme with the exception of technical issues.

The regular financial institutions could be considered as project developers if there is a real willingness to offer such services to their clients or to seize new market opportunities, although the real project developers emerge from the private sector and the investment firms. The latest knowledge is quite good, even if they do not have skills in the country; they have access to the group skills outside the country as it is the procedure. However the other project promoters admit not having the necessary skills to use this new financial scheme but usually request the services from professionals in such specific work. To reduce project development costs, they might need to have the basic skills for the first screening stage, before asking for professional services.

Private or public financial and investment institutions have enough capacities for equity and mezzanine financing development, but based on their interest, these skills were not available for the EE and RE project development.

Brief conclusions and recommendations

The assessment shows that there is a need for project development in order to provide the necessary tools for project realization for EE and RES projects. It is recommended to focus on the following issues:
- Increase human resources in the ministries dealing with the EE and RE projects and issues
- Capacity building for the ministries' resources and increase cooperation and collaboration between the different departments
- For the private sector, support for the project promoters for projects development and assistance for bankable projects realization
- Support for ESCO development.

Investor interest

Public sector

The Ministry of Economy's view of the Equity Investment Fund is quite positive and considered as reducing the limitation of public participation in the field of EE and RE investment. There are no legal obstacles for the State public organization to invest in the Fund and the fact that the ministry is in charge of the industrial sector might be a drive to invest in promoting EE in the designated sector. However, the lack of resources and skills within the ministry dealing with EE and RE reflect the absence of programmes and planning in that field and equally the misunderstanding about the Equity Fund and specificities among the other financing mechanisms.

Loss of control over invested resources has been mentioned in comparison with a national fund, over which the Government could keep full control. However, at the end of the privatization process where the Government will no longer be an owner, it could be interested in investing.

The Ministry of Economy and Regional Development, the Ministry of Mining and Energy, the Ministry of Environment and the Ministry of Finance stated that a potential public investment could only be decided by the Government through an interministerial agreement There is a will to draft a request and justification for the common interest in the activities related to EE and RE that will be considered by the government.

The governmental Environment Protection Fund (EPF) could be considered as a potential investor in the Fund. The director's proposals to provide participation for pilot projects implementation reflects EPF's interest. There are no legal barriers to such participation and the possible investment in the Fund is announced to be up to €3-4 million.

Private sector

In general, good interest has been noticed from stakeholders concerning the Fund. In the public and private sectors, mobilization for fund use and possible investment have been recorded, however the lack of stakeholders' capacities mitigates possibilities for real financial participation. There were also misunderstandings about how an equity and mezzanine fund would operate.

Potential partners for project co-financing

The financial institutions could be a potential partner for project co-financing and fund promotion. Despite the lack of activities related to EE and RE projects in the banking sector, the high potential for project developments needs close cooperation between the Fund and the financial institutions.

From the private sector, TEAM BRUNSWICK Group could be approached to investigate possible project co-financing for the project pipelines already identified by the Group.

Brief conclusions and recommendations

We can conclude that the public and private sector investors' interest in the Equity Investment Fund is medium. The main reasons reside in this period considered as a transition period characterizing the legal framework and regulation.

- Expressed interest from the mentioned stakeholders should be reiterated during the future project phases and followed up by the NPI, UNECE Project Management Unit and the Fund Designer.
- The Ministry of Mining and Energy should prepare a proposal for an interministerial discussion with the assistance of the UNECE secretariat.
- Laws on investment funds (made or to be made, in accordance with EU directives) and on public debt should be viewed with attention. Information can be requested from the Ministry of Finance.
- KfW has signed refinancing contracts with four Serbian banks of altogether €45 million for energy efficiency and renewable energy investments. The projects are being implemented since September 2008 (Cacanska Banka) and January 2009 (Volksbank, Raiffeisen, Pro Credit Leasing). The investments are aiming at SMEs and households. These banks should be approached as potential co-financing partners to the Fund.

THE FORMER YUGOSLAV REPUBLIC OF MACEDONIA

Energy Overview

The primary energy supply of the former Yugoslav Republic of Macedonia is dominated by coal (50 per cent) and crude oil (35 per cent). Around 10 per cent of the energy supply is based on renewable energy sources: hydro energy, firewood and geothermal energy. Almost 50 per cent of the primary energy supply is imported.

Industry is the largest energy consuming sector (30-35 per cent) with metallurgy (iron and steel industry) accounting for approximately 60 per cent of the total industrial energy use. The residential sector is the second largest energy consumer (30 per cent), and electricity is the major energy used in the sector (increasing, 53 per cent in 2005), including for space heating.

The restructuring of the electricity sector started in 2004 by unbundling the former vertically integrated state-owned power company ESM into four major companies. Power generation (ELEM) and transmission (MEPSO) will remain state owned; the distribution company (ESM) is 90 per cent owned by the EVN AG (Austrian power distribution utility); the authorities proceeded with the sale of 100 per cent stake in the one-plant generation company (TEC Negotino), but later on decided to cancel the sale and keep the plant under state ownership.

The Government of the former Yugoslav Republic of Macedonia gives priority to the development of renewable energy sources, as the country has promising resources of renewable energy, including hydropower, geothermal energy, biomass, and wind energy.

The Government considers the construction of small hydropower plants as one of the projects of great importance for the country. An international public competition was held in 2006-2007 for granting water concession for electricity generation from 60 small hydropower plants according to the DBOT model (Design, Build, Operate, Transfer).

One major problematic area in the country is the wide use of electricity for domestic heating and the inefficient energy consumption in buildings. The Government has started addressing these problems, but they have not been given enough priority. The introduction of a building certificate system is planned.

Currently, less than 9 per cent of the households in the former Yugoslav Republic of Macedonia are connected to district heating networks. Five district heating systems are operational with total capacity of 653 MW, powered by heavy oil (more than 75 per cent), natural gas and lignite. The biggest district heating system with a capacity of 518 MW is serviced by "Toplifikacija" AD – Skopje (a private company). However, there is no equipment for individual metering and control of heat consumption in collective buildings and the bills are calculated on the basis of square metres of floor area. Expansion of the district heating systems is not foreseen in the short term.

The existing building stock in the former Yugoslav Republic of Macedonia is not energy efficient. In general, the level of thermal insulation of the buildings is poor. Old buildings which are constructed using traditional methods with thick brick or stonewalls can offer a rather acceptable comfort level. The main problem is encountered in new constructions of reinforced concrete, where the heat losses are excessive and the comfort level is low.

A regulation for thermal insulation of buildings has been in force since the 1980s, but its actual implementation is limited. Building permits are received without any energy efficiency criteria. There is no enforced energy performance standard. There is no supervision of the energy systems in buildings during their construction.

The considerable potential for improving energy efficiency in buildings, together with the high level of electricity use by households, makes the residential sector a key target for energy efficiency improvement programmes. Priority targets for commercial projects include lighting systems, space heating and hot water.

Financial environment in energy efficiency and renewable energy

Legal, regulatory and policy framework

The strategic priorities in the energy sector are incorporated in the Energy Law, adopted by the Parliament in 2006. Provisions for energy efficiency are included in the Energy Law. There are on-going efforts for developing and adopting the secondary legislation and technical regulations, as only the labelling of household appliances is regulated so far.

The Energy Law allocates the responsibilities for energy efficiency policy development and implementation in the former Yugoslav Republic of Macedonia to the Ministry of Economy, supported by the Energy Agency. The Energy Law has involved local authorities in energy efficiency strategic planning and programmes implementation.

The objective of EU accession heavily influences the development of the energy policy in the former Yugoslav Republic of Macedonia.

The former Yugoslav Republic of Macedonia is structured in 84 municipalities and the city of Skopje as a separate unit of local self-government. The regional reform is going ahead with fiscal decentralization in place since July 2005. Following two years grace period, the municipalities will gain new responsibility to manage their future investment needs with their revenues. In the new legislation, municipalities will be able to borrow directly from IFIs and to offer financial guarantees in respect of loans to their municipal utility companies.

The Energy Law obliges municipalities and the City of Skopje to develop and implement five-year local energy efficiency programmes and action plans for their implementation. The Norwegian Government has provided financing to a capacity building programme supporting four municipalities in preparing their energy efficiency plans.

A new energy strategy on renewables should be finished by January 2009, and the municipalities will need to adjust their plans accordingly.

National priority areas for EE and RES

The energy development in the former Yugoslav Republic of Macedonia has major implications for national economic growth, environmental protection and human livelihood.

Funding is one of the critical steps in establishing successful energy programmes.

One of the main national priorities is to establish an energy efficiency fund. The fund should be maintained and operated outside the government by commercial banks. The fund could be utilized as direct loans, or could be used to provide guarantees on loans issued by the commercial banks with their own capital. Other national priorities could be improvement of institutions and administrative capacity, introducing a system of certified energy auditors, building energy code and standardization of equipment.

Initiatives in implementation of a series of technical programmes should be treated as national priorities particularly in the field of residential buildings, commercial building, industry programmes and street lighting programme.

Interest in receiving equity and mezzanine financing

The Ministry of Economy, responsible for EE and RES, showed strong interest in the Fund, which could provide good opportunities for implementing several EE and RES projects.

According to ELEM, the state owned electricity generation company, the legal framework is in place for establishing wind farms as private companies, and the Fund would be an ideal partner.

Small Enterprise Assistance Funds (SEAF) (Fund Management Company) has made 14 investments until now in the former Yugoslav Republic of Macedonia, and is still active in four of them. SEAF is interested in syndicating new projects with the Fund.

Regarding the new Fund, it was stated to be important that the equity provided could be used as financing in projects.

Barriers to financing EE and RES

There are several barriers to realization of EE and RES projects:
- Lack of capacities and skills to utilize the available funds in the former Yugoslav Republic of Macedonia, including how to prepare project proposals properly
- Existing legal framework (including public tendering, concession, etc)
- Overall electricity price levels are low
- Lack of public awareness
- Lack of sufficient capacities to deal with energy efficiency in municipalities
- Political risk (replacements after elections, including at municipal level)
- To provide guarantees – money for lending is available
- Municipalities not allowed to take loan (as of today, only two municipalities can take loans, based on guarantee provided by the Ministry of Finance).

International tenders have been organized for a large number of small hydropower plants, several offers received and winners selected. However, yet no contract has been signed. The Government is still discussing how this can be organized properly. Concession can not be given for state owned land, leasing is possible. At some sites the Government would need to expropriate privately owned land, register the land as state owned, and then do leasing. Leasing period = 20 years (plus three years for construction). For wind farms, the leasing period could be more than 20 years. Another barrier is the environmental impact.

According to Toplifikacija, there is interest in energy efficiency projects in the private sector. They expect that the market within municipalities could be more interesting when the municipalities get into the second phase of decentralization. Only then they will be allowed to take loans. However, the rules for public tendering might be a barrier. Another barrier is that there are no departments responsible for energy efficiency, as well as the lack of awareness of energy efficiency among municipal decision-makers. Changing of staff every four years is also a barrier.

Toplifikacija have not started to look into RES business yet because of unclear rules and regulations. They expect that the new Strategy on Renewables from January 2009 will facilitate RES business.

To address some of the existing barriers to energy efficiency, the Government of the former Yugoslav Republic of Macedonia initiated the Sustainable Energy Project, supported by GEF and the World Bank. The project is aimed at introducing ESCOs and loan/guarantee facility as two instruments for financing energy efficiency and small scale renewable initiatives.

Another important incentive is the establishment of feed-in tariffs for renewables.

In February 2007, the Rulebook on feed-in tariffs for purchase of electricity produced from small hydropower plants was published by the Energy Regulatory Commission (ERC). The feed-in tariffs apply to the quantity of electricity produced and delivered by newly constructed run-of-river small hydropower plants, which have qualified as privileged producers (see table). The privileged producer is obliged to use the feed-in tariffs approved for him for 20 years. The electricity market operator is obliged to purchase the total quantity of electricity delivered by the privileged producer under the approved feed-in tariffs.

Feed-in tariffs for the sale of electricity produced by small hydropower plants

Delivered quantity (block)	Monthly quantities of delivered electricity (kWh)	Annual quantities of delivered electricity (kWh)	Privileged tariff (€cents/kWh)
I	1 – 85 000	1 – 1 020 000	12.0
II	85 001 – 170 000	1 020 000 – 2 040 000	8.0
III	170 001 – 350 000	2 040 001 – 4 200 000	6.0
IV	350 001 – 700 000	4 200 001 – 8 400 000	5.0
V	Above 700 000	Above 8 400 001	4.5

(Average price for the production of 1,000,000 kWh per month will be 58,800/ 1,000,000 = 5.88 euro cents/kWh)

The Rulebook on the method and procedure for determination and approving the use of feed-in tariff for purchase of electricity generated by wind power plants was adopted by ERC in May 2007 (8.9 euro cents/kWh, ex. VAT).

The Rulebook for feed -n tariff for purchase of electricity produced by power facilities using biogas as fuel was adopted by ERC in November 2007 (13 euro cents/kWh, ex. VAT for biogas facilities with installed capacity ≤ 500 kW or 11 euro cents/kWh, ex. VAT for biogas facilities with installed capacity >500 kW).

The Rulebook for feed-in tariff for purchase of electricity produced by PV power facilities was adopted by ERC in September 2008 (46 euro cents/kWh, ex. VAT for PV facilities with installed capacity ≤50 kW or 41 euro cents/kWh, ex. VAT for PV facilities with installed capacity >50 kW).

A 13.5 per cent increase in electricity tariffs has recently been approved.

According to ERC, cross subsidies do exist. Tariffs should be cost +. The electricity market should be open from 1 January 2015.

A credit and guarantee scheme is established through the Macedonian Bank for Development Promotion and five local, commercial banks: Komercijalna Banka, Uni Banka, NLB Tutunska Banka, IK Banka, and Ohridska Banka.

EBRD:

- Western Balkans Sustainable Energy Direct Financing Facility. € 50 million to provide direct individual loans between €1 and €6 million to industrial energy efficiency and renewable energy projects. The facility is complemented with grant funding for technical assistance for project identification, preparation and implementation verification.
- Western Balkans Sustainable Energy Credit Line Facility. € 50 million to provide loans to participating banks for on-lending to sub-borrowers for energy efficiency and renewable energy investments up to €2 million per project in the industrial sector and € 200,000 in the residential sector. The facility is complemented with grant funding for marketing and awareness raising, establish technical eligibility criteria, prepare/appraise projects, implementation verification, etc.

USAID Macedonia has introduced two Development Credit Authority facilities to support SMEs and energy efficiency projects in municipalities. The SMB facility is $ 9 million split between two financial institutions - Unibanka ($ 5 million) and NLB Leasing ($ 4 million). The energy efficiency facility is $ 10 million split equally between the same two financial institutions. The first disbursements started early 2008.

The credit market for municipalities is non-existent, and USAID believe that by providing this facility strong and credible relations between the private lenders and the local government units would be established and developed.

In addition, new and existing SMEs need finance to expand their operations and enhance their growth. USAID also believe that by this guarantee that covers 50 per cent of the banks risk, banks would engage in more non-traditional lending activities.

The Ministry of Finance and Association of Local Self Government Units fully supports this initiative.

A joint office between the Ministry for the Environment, Land and Sea of Italy and the Ministry of Environment and Physical Planning of the former Yugoslav Republic of Macedonia was established in 2005. There are similar offices in other Balkan countries (e.g. Croatia). Their main aim is to identify and propose CDM project ideas, for which Italian and Macedonian companies offer to develop a project document design (PDD).

One project, rehabilitation of six small hydropower plants, was not approved due to lack of sufficient additionallity. PDD's for 10 biomass co-generation plants are now being evaluated. A PIN has been prepared for a 40 MW_e gas fired CHP plant.

SEAF is a global investment firm focused on providing growth capital and operational support to businesses in emerging markets and those underserved by traditional sources of capital. SEAF selectively makes structured debt and equity investments in locally owned enterprises with high growth potential.

Based in Washington D.C., SEAF invests in more than 30 countries around the world through an international network of 19 offices in Central and Eastern Europe, Latin America, and Asia. Investors include a cross section of public and private institutions, including several of the international finance institutions, local pension funds, insurance companies, banks and family offices.

As a global organization with a local presence, SEAF is able to provide management with hands-on operational support and provide businesses in emerging economies with the global "connections" that accelerate their growth and profitability.

SEAF began its career of fund management in Europe when, in 1992, it established its first fund in Poland: CARESBAC Polska. Since then, SEAF has opened numerous fund offices throughout Central and Eastern Europe and has investments in Poland, Bulgaria, Croatia, Romania, the former Yugoslav Republic of Macedonia, Estonia, Latvia, Lithuania, Georgia, Serbia and Montenegro. SEAF's history in Europe includes many exited investments and five retired funds.

KFW: There is no information on KfW in the former Yugoslav Republic of Macedonia, but the representative in Albania is also responsible for the former Yugoslav Republic of Macedonia.

The World Bank has announced the establishment of a credit line that should become operational in spring 2009. The total credit line will be about $25 million and will support municipal infrastructure projects in the former Yugoslav Republic of Macedonia, including energy efficiency projects. The interest rate shall be based on LIBOR. The payment of loans will be up to a maximum of 17 years with 3 to 5 years of grace period.

ESCOs

The MT ESCO established within the GEF/World Bank Sustainable Energy Project, as a joint venture between MEPSO (state owned transmission company) and Toplifikacija (private owned district heating company in Skopje), is still not operational. A new Memorandum of Understanding regarding operational support and funding of the ESCO is being prepared, and needs to be approved by the World Bank. The plan includes replacing MEPSO with ELEM (the state owned power generation company). It is expected that the ESCO staff will be operational three months after the Memorandum of Understanding is signed , and they will be starting from scratch.

Toplifikacija Engineering was established in 2007. In addition to providing engineering services to the mother company, they also provide services to other clients. Their field of business includes:
- design and documentation of central and district heating systems
- design and documentation for HVAC (heating, ventilating and air conditioning) systems
- energy efficiency projects
- trade of HVAC equipment

Due to the lack of progress with the development of MT ESCO, Toplifikacija decided to offer EE services through their engineering company, and the first "ESCO-type" project was planned to start up by the end of 2008, energy efficiency in a hotel. A loan was provided by the Macedonian Bank for Development Promotion.

The existing FONKO ESCO has not yet completed any real, full scale ESCO project.

According to the Ministry of Economy, it would be useful to have several ESCOs in operation to ensure sufficient competition.

Banking sector

No meetings were organized with the banking sector.

SEAF has made 14 investments so far in the former Yugoslav Republic of Macedonia, and is still active in four of them. The return for the investors has been quite good. SEAF has gained a lot of experience that should be of interest for the Fund Designer and the Fund Manager, both in the former Yugoslav Republic of Macedonia and other countries in the region. SEAF is interested in syndicating new projects with the Fund.

Brief conclusions and recommendations

- The legal and regulatory framework for investments in RES seems to be unclear, the Government is still discussing how to arrange leasing of sites for small hydropower plants and wind farms
- Feed-in tariffs for electricity production from RES is in place
- After having the necessary legal and regulatory framework in place, equity participation in the Small Hydro Power sector seems to be the most promising (the adoption of a Strategy for Renewable Energy Sources is foreseen early 2009)
- The Energy Law obliges municipalities to develop and implement five-year local energy efficiency programmes and action plans for their implementation. This should create a larger market for energy efficiency in the former Yugoslav Republic of Macedonia;
- When the second stage of decentralization is completed, municipalities will be able to borrow directly from IFIs and to offer financial guarantees in respect of loans to their municipal utility companies;
- There are several barriers to energy efficiency in municipalities; including rules for public tendering, no departments responsible for energy efficiency, lack of awareness of energy efficiency among municipal decision-makers. Changing of staff every four years is also a barrier;
- If the ESCO within the GEF/World Bank project starts its operation within a short period of time, this might be an interesting possibility for the Fund;
- Toplifikacija (private owned district heating company in Skopje) could be an interesting partner for the new Fund;
- It would be useful for the Fund Designer and later the Fund Management to learn about the experiences from SEAF, local operations and its headquarters.

Energy efficiency and renewable energy sources project development and finance capacities

Existing and prospective EE and RES projects

ELEM, the state owned electricity generation company, has done some EE projects in their generation facilities. A feasibility study for a 50 MW wind farm, supported by the European Reconstruction Fund, should have been finished by April 2009, after which they would start seeking financing.

Toplifikacija is working on a modernization programme for their district heating system, including sub-station renovation. They have started to build one large co-generation plant, and two smaller ones (with Japanese support). One power plant is built with Russian co-financing. The company is also involved in some geothermal projects in the eastern part of the country. The projects are financed with a mix of equity and loans from local and international financial institutions.

Within the USAID Development Credit Authority a $630,000 loan has been provided for an energy efficiency school project in Karpos municipality by Unibanka.

Based on the meetings arranged by the NC and NPI, it is the impression that the level of capacities and skills operationally available for the development and financing of energy efficiency and renewable energy investments in the former Yugoslav Republic of Macedonia is low. On specific questions, all we met with confirmation that there is such a need for capacity building.

SEAF proposed to contact its headquarters in Washington for possible cooperation on capacity building activities.

Public sector

According the Ministry of Economy, there is not enough capacities and skills to utilize the available funds in the former Yugoslav Republic of Macedonia, including how to prepare project proposals properly. This is the case for national (ministries, committees and agencies, including the Energy Agency (National Participating Institution)), regional and local governmental bodies and state owned organizations. Lack of public awareness is another challenge.

Private sector

There are some experts that seem to have capacities and skills that could be utilized for the new Fund, but the resources are too limited to support a wider market development.

There s no evidence that FONKO ESCO has completed any real, full scale ESCO project. Since the planned MT ESCO is not yet operational, they have no experience with this type of operation. During the meeting with them, they stated that international support would be requested to support the development of the ESCO.

Regarding energy efficiency in public and private buildings, we only learned about some very few demonstration projects. Without a number of projects being realized, the capacities and skills on development of bankable energy efficiency projects is most likely limited.

During the meeting with Toplifikacija Engineering, they confirmed that more capacity building would be needed. As of today, only three persons are involved in the energy efficiency activities.

The local World Bank consultant team for the GEF/World Bank Sustainable Energy Project had some experience in providing technical due diligence to World Bank projects. The financial due diligence was done by the local banks. According to the World Bank local consultant, financial engineering capacity building (preparation of business plans) is urgently needed. To ensure sustainability, training of trainers should be prioritized. Furthermore, risk management is an important issue.

NGOs

Further capacity building seems to be beneficial also for the NGOs we met during the mission.

The feasibility studies were prepared for a number of SHP projects (and one wind farm project), however there is no information on the quality of any study. As we did not learn about any RES project being implemented (nor decision taken to do so), this could also indicate that the capacities and skills available in this field are limited.

There is no information on any special purpose companies established for instance for SHP projects, knowledge and skills on equity and mezzanine finance to be very limited.

Brief conclusions and recommendations

- There is a substantial need for capacity building in the former Yugoslav Republic of Macedonia, both related to investment project development and equity and mezzanine finance business development. There are some experts that seem to have capacities and skills that could be utilized for the new Fund, but the resources are too limited to support a wider market development;
- Capacity building is also needed to strengthen the various ministries, committees and agencies (including the Energy Agency) involved in the field of EE and RES, and better coordination and cooperation between them would be helpful.

Investor interest

Public sector

The issue of potential participation in the Eastern Europe Energy Efficiency Investment Fund was presented and briefly discussed with some parties during the mission. No public sector interest can be reported. However, it is important to bear in mind that the mission did not learn many real candidates for such a potential interest.

Private sector

No private sector interest for participation in the Fund can be reported. However, it is important to bear in mind that the mission did not meet with many real candidates for such a potential interest.

The interest from IFIs needs to be clarified by their headquarters, and not through the branch offices in the former Yugoslav Republic of Macedonia.

Potential partners for project co-financing

SEAF is interested in syndicating new projects with the Fund.

Brief conclusions and recommendations

No real interest from public and private investors in the former Yugoslav Republic of Macedonia in participating in the European Energy Efficiency Investment Fund was identified. The interest from international finance institutions needs to be clarified by their headquarters, and not through the branch offices in the former Yugoslav Republic of Macedonia.

UKRAINE

Energy overview

Existing energy resources, energy dependence for primary and secondary energy resources, production of electricity and of heat, use of renewable energy sources (RES)

The energy sector of Ukraine is based largely on oil and gas, of which approximately 75 per cent is imported. These two energy sources account for 55 per cent of the total primary energy supply (TPES). Domestic extraction of crude oil is about 4 million tons per year, and domestic production of natural gas is about 20 billion cubic metres per year. Coal, which is mainly a domestic source, accounts for 31 per cent. Its annual output is around 80 million tons. The remaining 14 per cent of TPES consists of nuclear energy (13 per cent) and renewable sources (1 per cent). One of the declared priorities of the Government is making the country less dependent on oil and gas imports.

Ukraine's electricity sector has about 54 GW of installed capacity. The bulk of it is thermal power plants and combined heat and power plants (close to 60 per cent) and nuclear power plants (NPP) (25 per cent). However, thermal power plants are often running at low capacity and produce less electricity than NPPs. 96 per cent of thermal power plants have reached the end of their service life, with almost half of them having exceeded their maximum service life. The share of NPPs (15 operating power units) in electricity generation is close to 50 per cent.

The heat sector is dominated by district heating. The total length of the pipelines is 45,000 km, and the total capacity of the network is 200,000 MW of heat. However, district heating distribution networks in Ukraine are outdated, sometimes poorly insulated, and losses are significant. Multi-story residential buildings consume approximately 40 per cent of the country's heat energy resources. It is estimated that fuel consumption in the heat sector could be reduced by up to 30 per cent simply by improving equipment such as boilers, pipes, pumps and valves. Further energy savings might be obtained through appropriate design of plants and effective metering of heat consumption in the household sector.

Although the energy intensity of the Ukrainian economy has been decreasing in recent years, it is still one of the least energy-efficient countries in the world. Factors that have contributed (and still contribute) to the high energy intensity include slow restructuring of energy-intensive industries; old capital stock in the public, private and household sectors; and inadequate reforms of the heat and power sectors. The most important factors are low tariffs in the heat and power sectors and the prevailing cross-subsidization of households by industrial consumers. Generation facilities converting primary energy into heat and power have low efficiency rates, and technical and commercial losses in the transmission and distribution networks are high.

Even though Ukraine has strong potential for use of RES and has adopted a number of laws and policies promoting it, the share of RES remains low (around 3 per cent).

Level of priority given to EE and to RES in country's energy policy

Ukraine has a programme of state support for the development of non-traditional and renewable energy sources and small hydropower plants. The target set for renewables is 10 per cent of power generation by 2010 but it is not likely to be achieved.

The most developed RES for electricity production in the country are wind turbines (total installed capacity about 200 MW) and small hydropower plants (total installed capacity about 105 MW). Mid-term potential for wind generation is estimated at 5,000 MW, with the Crimean Peninsula the most promising area for installation of wind farms. A special "green" tariff for wind energy has been established but is applied only to domestically produced wind turbines. It is not clear whether it is being applied in practice. Another obstacle is lack of infrastructure (transmission lines) to bring the electricity from wind farms into the system.

Meeting at the Ukrainian Union of Industrialists and Entrepreneurs

Ukraine has potential for solar energy, particularly in Crimea and in southern Ukraine. Water heating with solar energy is getting particular emphasis. Potential for small hydropower projects is estimated at about 330 MW, of which 220 MW on the Tisa River in Western Ukraine.

The country has considerable geothermal resources that are used primarily for heat supply. Total installed capacity of thermal systems is 13 MW_{th}. Plans are in place to increase the thermal water utilization up to 250 MW_{th} by 2010. There are prospects for binary geothermal plants using existing wells at abandoned oil and gas fields.

The biomass potential is estimated at 4 million toe and includes livestock manure, straw, and timber waste. There is some interest in the use of livestock manure for biogas power generation as well as straw and wood combustion for district heating plants and combined heat and power facilities. Some Ukrainian companies have begun production of rapeseed oil and its export to European biodiesel producers. There are plans to build biodiesel production plants with the participation of foreign investors.

Investments in the energy sector

Significant investments are made in the energy sector of Ukraine. In 2008, about $1 billion capital investments were made in the power industry, which is 17.3 per cent more than in 2007. Major capital investments projects included construction of Dniester and Tashlyk pumped storage hydropower stations; renovation of major turbine generators and a chain of hydropower plants on the Dnipro River; and construction of electricity transmission networks from nuclear power plants.

Enterprises of the National Joint Stock Company Naftogas of Ukraine made capital investments of over $0.7 billion in 2008, about 0.4 per cent more than in 2007. The main investments were in putting on line natural gas and oil wells and in the construction of natural gas and oil pipelines. State coal companies invested about $0.2 billion from all funding sources, which is 11 per cent higher than in 2007. Investments were intended for construction of new coal mines, reconstruction of operating ones and technical refurbishment of coal enterprises.

The Agreement signed by the Ukrainian Government and EBRD in 2007 opens up opportunities of increasing investment in the public sector of Ukraine (national and municipal levels) in the next three years. EBRD funds are intended for various sectors of Ukraine's economy, including energy industry and energy efficiency improvement. The EBRD investments in the Ukrainian public sector are expected to reach €400 million in 2009. The projects under the Agreement include construction of high-voltage power transmission lines in several regions, installation of efficient natural gas compressor stations, renovation of underground natural gas storage facilities, and improvement of municipal infrastructure, such as district heating and wastewater facilities.

Financial environment in energy efficiency and renewable energy

Legal, regulatory and policy framework

Major government institutions involved in developing and implementing policies and actions in the fields of energy, energy efficiency, and renewable energy sources include the following:

On energy:
- Ministry of Fuel and Energy
- Ministry of Coal Industry
- National Electricity Regulation Commission (NERC)

On energy efficiency:
- National Agency for Efficient Use of Energy Resources

On renewable energy sources:
- National Agency for Efficient Use of Energy Resources

The National Agency for Efficient Use of Energy Resources was been established at the end of 2005 after the State Committee on Energy Saving had been abolished. It is responsible for national policy in the area of energy efficiency, energy conservation and development of alternative energy sources. In addition, it works directly with companies. The Agency recommends companies that would like to implement energy efficiency projects for investments from the budget (annually about UAH 600 million (about $120 million at the exchange rate at the time of the mission)). The Agency remains a relatively weak body with a limited budget.

The Ministry of Environmental Protection is responsible for policies and activities related to climate change mitigation.

The Ministry of Housing and Municipal Economy is responsible for policies and activities related to the residential sector, including increasing energy efficiency in buildings, improvements in heat supply, etc.

Legislation in the areas of energy and energy efficiency

Ukraine has a complicated legal framework for the energy sector, including energy efficiency and energy saving. The Law on Energy Saving (1994) provides for a system of institutional and regulatory measures and incentives for energy saving. Other main legislation includes the Law on Electricity Sector (1997), the Law on Heat Supply (2005), and the Law on Combined Heat and Power Production (Cogeneration) and Utilization of Energy Waste Potential (2005). The laws are supplemented by a number of government resolutions, Presidential decrees, by-laws, regulations, norms, standards and methodological guidelines.

A number of legislative documents are under development. In particular, a draft law on energy efficiency in buildings has been developed, along with a package of draft norms and legislation on reforming housing and municipal economy.

Policy documents in the areas of energy and energy efficiency

The main energy policy document is the Energy Strategy of Ukraine for the period up to 2030 (2005), which replaced the National Energy Programme up to 2010 (1996). The Strategy focuses on traditional energy sectors (natural gas, oil, nuclear and coal). One of its goals is reducing the country's energy dependence, in particular on natural gas imports. The goal of increasing the share of alternative energy sources in fuel and energy balance to 19 per cent is stated, however it is not clear how it could be achieved.

The main policy document in the area of energy saving remains the Comprehensive State Programme of Energy Saving of Ukraine (1997). Many of the targets established by the Programme have been achieved (by some estimates around 90 per cent), therefore a new programme or strategy on energy efficiency and energy saving could be useful. Issues related to energy saving and energy efficiency in housing and communal sector are reflected in the State Programme of Reform and Development of the Housing and Communal Sector for 2004–2010 (2004).

A National Strategy on heat supply is being developed.

Legislation and policy documents in the area of renewable energy sources

Several laws and policy documents have been adopted to promote the use of renewable energy sources. They include the Law on Alternative Types of Liquid and Gaseous Fuel (2000), the Law on Alternative Energy Sources (2003), the Programme of State Support of Development of Alternative and Renewable Energy Sources and Small Hydropower and Thermal Power (1997, modified in 2005), the Concept (Outline) of the Programme of Development of Diesel Biofuel Production for the period up to 2010 (2005), and the Resolution of the Cabinet of Ministers on Further Development of Wind Power in Ukraine (2007).

The structure of the energy market and the main players

Electricity sector

Ukraine's Wholesale Electricity Market began operating in 1997. The state-owned company Energorynok operates the market, serving as a single buyer of electricity. In principle, the large thermal power generation companies may compete to sell power to Energorynok. Energoatom and Ukrhydroenergo also sell power (respectively from nuclear and hydropower plants) to Energorynok, but at prices set by the National Electricity

Regulatory Commission (NERC). Thus, the competitive wholesale supply accounts for only about 35-40 per cent of the power sold to Energorynok. Energorynok then sells power to the Oblenergos (regional electricity distribution companies) and large industrial companies. NERC sets the regulated prices for transmission and distribution services. In most cases, prices are set based on cost recovery. In turn, the Oblenergos sell power to customers at rates that are based on the wholesale price plus the transmission and distribution tariff. Overall, the market for electricity prices is very limited, and competition is almost inexistent. NERC also issues licenses, regulates activities and tariffs for heat from combined heat and power plants and renewable energy sources.

The main player in the energy market is the state-owned national joint stock company Energy Company of Ukraine (NAK EKU). The justification for its creation was to ensure the economic and energy security of the state and an efficient functioning of the power sector. NAK EKU owns state holdings of shares of 43 power enterprises of Ukraine, including the state holdings of shares of the national joint stock company "Ukrhydroenergo". The charter capital of NAK EKU is about UAH 10 billion. The only private company operating in the market is Vostokenergo.

Coal sector

Ukraine's coal market capacity is estimated at $6.0 billion per year. Annual coal production is about 80 million tons. State-owned mines still account for over 90 per cent of total coal output.

Oil and gas sector

Naftogas of Ukraine and its constituent entities account for about 94 per cent of total natural gas production and 97 per cent of total oil and condensate production. Domestically produced natural gas is distributed mainly to households and public buildings (hospitals, schools, etc.) at subsidized prices.

The Ukrainian gas transmission and storage system consists of about 37,600 km of gas pipelines and 13 underground gas storages with a total capacity of over 33 billion m^3. When fully loaded, the underground gas storages can deliver up to 240 million m^3 per day.

National priority areas for EE and RES

As most of the industrial enterprises are privatized, improvement in their energy efficiency is implemented along with the general modernization and improvement of technologies. Energy efficiency improvements in industry are expected to be part of the overall investment process in developing the industrial sector of Ukraine driven by the need to reduce expenses for electricity, gas and water. Until recently, rising energy prices provided a strong incentive for these improvements.

Various policy documents set targets for the share of renewables in energy production. The Programme of State Support of Development of Alternative and Renewable Energy Sources and Small Hydropower and Thermal Power Ukraine sets this target at 10 per cent by 2010. However, at this point it is not realistic to expect this target to be achieved.

The main national priority areas for renewable energy sources identified in various policy documents and legislation include wind, solar, geothermal, biomass and hydropower. However, actual support for their development by the Government is far from sufficient and remains mainly declarative. The potential for RES in Ukraine is described above.

There are a number of issues affecting the interest, capacity, willingness and possibility to invest in renewable energy and energy efficiency projects in the current financial climate.

- The legal and fiscal mechanisms for converting the savings from increased energy efficiency into revenue which can be used to service loans are not yet fully developed, despite the existence of the respective framework legislation.
- The borrowing capacity of municipalities is low and is not expected to increase in the short term. Collection of money from households, affected by the economic crisis, for improvements of private dwellings would be problematic.
- Most of the rehabilitation projects are relatively small – from $100,000 to $2 million. However, in some cities, including Kharkiv, Zaporizhzhya, Lviv and Rivne, there are opportunities for major projects estimated at $10-15 million or more. Some of the projects, such as renovation of combined heat and power plants are of a much bigger scale ($100-200 million).
- The financial crisis has already had a negative impact on the economy of Ukraine. In these circumstances, energy savings will be even more important, but the lack of available credit in the market and the falling energy prices may make investment in energy efficiency less attractive. Availability of budget financing for energy efficiency and renewable energy may decrease further.

Barriers to financing EE and RES

Despite the declared interest in investments, there are a number of barriers facing any potential investor and a perception of risk which makes investments in this sector difficult, particularly for smaller investors. Some of the main issues include:

- Subsidized nature of domestic tariffs for electricity and heat, i.e. the tariffs are set by a government agency and can be set arbitrarily to meet current political objectives, with the interests of investors not taken into account.
- Difficulties in monetizing achieved savings – for example, when savings are achieved by an enterprise or organization that receives funding from the national or local budget, the money saved will not be disbursed to the entity and will remain in the Government's budget, thus it cannot be used for repaying the investment.
- Investing in projects owned by state and municipal entities always raises the issues of guarantees and collateral: can the project be guaranteed by the state (slow and cumbersome process, often not practical for projects less than $50-100 million), what is the value of such guarantee if the credit rating of a particular country is low, can a municipal or public property be accepted as a collateral and many other similar legal and financial issues related to that.

There are major policy, legal and institutional obstacles, including:

- Absence of the consistent long-term state policy in the field of energy efficiency
- Many of the measures stipulated in the relevant state programmes remain on paper only
- Objectives of the state programmes often lack practical measures due to unavailability of resources
- Inefficient system of administration in the area of energy saving with frequent changes and lack of institutional capacity
- Imperfect regulatory and legal framework that hinders the efficient management of the energy saving process, implementation of the measures stipulated by the

corresponding programmes and attraction of the available state resources.

Technical obstacles include the following:
- Insufficient information and equipment for measuring energy consumption
- Lack of necessary skills in the companies to develop bankable investment proposals and business plans
- Lack of public awareness regarding energy efficiency benefits and opportunities (energy saving is still low priority to many users as a result of long period of cheap energy prices)
- Lack of training programmes to better inform businesses and other energy consumers of how to improve energy efficiency
- Lack of the infrastructure for energy audits
- Slow progress in updating building energy codes
- Lack of energy efficiency standards for equipment

Economic and financial obstacles, including:
- Lack of the owner's interest (particularly in public entities) in the reduction of the production expenditures
- Monopolistic nature of the production markets
- Difficult access to credit resources due to the banks' reluctance to give loans for investments with more than a one year payback period
- Relative weakness of the banking system does not allow financing large-scale energy saving projects
- Too high requirements of the banks regarding the liquidity and amount of collateral
- High level of inflation
- Distorted energy pricing that does not reflect full, long-term economic cost.

Incentives

There are economic instruments to provide incentives for energy efficiency and use of renewable energy sources, including investment incentives.

In 2007, the Parliament adopted the Law on Amending Some Legislative Acts of Ukraine Regarding the Encouragement of the Energy Saving Measures, which envisages a series of tax incentives in the area of energy efficiency. Specifically the taxes are not charged from:
- Profit of the enterprises obtained from selling of the locally produced goods at the custom territory of Ukraine based on the list approved by the Cabinet of Ministers of Ukraine: equipment that uses alternative and renewable energy sources; energy saving equipment and materials and goods that provide for saving and rational use of the fuel and energy resources; tools for metering, control and management of the fuel and energy consumption; equipment for the production of alternative fuel.
- Profit of the enterprises included in the State register of the enterprises, institutions and organizations dealing with development, implementation and use of energy saving measures and energy efficiency projects but overall not more than 50 per cent of taxable profit. The state register includes the enterprises that are included in the sectoral energy saving programmes based on the results of the expertise (assessment) conducted by the State Inspectorate on energy conservation and received conclusion that their energy saving measures and energy efficiency projects meet the energy saving criteria and are included in the sectoral energy saving programmes.

In April 2008, the Cabinet of Ministers adopted a Resolution on the procedure of using the money allocated in the 2008 State Budget for the state support of energy saving measures. This procedure specifies the mechanism of using the money allocated by the 2008 State Budget for the programme "State support of the energy saving measures through the mechanism of the reduced loan interests". Budget money is used for the offset of the interest rate for the loans attracted by the entities for financing of the energy saving measures, including energy efficiency projects.

Financing schemes

Sources of financing for energy efficiency and renewable energy projects

Government sources

Government sources of financing include the State (national) Budget and local budgets at all levels. In 2005-2006 the Government (including local budgets) spent about UAH 800 million (approximately $60 million) to implement measures envisaged by the Comprehensive State Programme of Energy Saving of Ukraine and the Programme of State Support of Development of Alternative and Renewable Energy Sources and Small Hydropower and Thermal Power.

In early 2008, the State Energy Conservation Fund was established as a budgetary fund for financing energy efficiency improvements.

Businesses

There is no detailed information regarding private domestic investment companies or other business entities specialized in financing energy efficiency and renewable projects. Nevertheless many Ukrainian companies (both state-owned and private) invest in such projects. According to a survey done by the State Control and Revision Service, about UAH 4.8 billion (about $0.9 billion) were invested by Ukrainian companies for implementation of projects included in the two state programmes mentioned above.

Some foreign private companies, such as Global Carbon, are known to be acting as brokers for GHG emission reductions in the country.

International and foreign organizations

The World Bank cooperates with Ukraine in the framework of several mechanisms, including Carbon Partnership Facility (with capitalization of $5 billion) with its two structural units: Carbon Assets Development Fund and Carbon Fund, as well as climate investment funds (Strategic Climate Fund and Clean Technology Fund).

EBRD supports energy efficiency investments in Ukraine. It intends to implement the Ukraine Energy Efficiency Programme-2 (UKEEP-2) following the success of the already existing €100 million UKEEP programme for energy efficiency and renewable energy projects in Ukraine (up to $5 million). EBRD is implementing Sustainable Energy Initiative, which includes areas of energy efficiency, renewable energy sources, and carbon financing. Total amount of EBRD investments in Ukraine is around $1 billion, of which up to 20-30 per cent could be in the energy efficiency projects.

Six main areas of EBRD activities:
- Energy efficiency in industry
- Sustainable Energy Financing Facility – credit lines for Ukrainian banks (for small projects up to € 5 million). In addition, the Technical Cooperation Fund provides training for bank managers and company managers and for energy audits of companies
- Energy efficiency in the energy sector and gas sector

- Renewable energy
- Energy efficiency in the municipal sector (focused on heat supply)
- Carbon finance – Multilateral Carbon Credit Fund.

Kharkiv Combined Heat and Power Plant (CHP-3)

USAID approved financing ($ 1 million) for implementation of energy efficiency projects in industry.

Other carbon funds and institutions are engaged in Ukraine and buy emission reduction units from projects that are suitable for JI (NEFCO, NEFCO Carbon Fund, ERUPT, DEPA, KfW Carbon Fund, etc.) At the time of the mission, Ukraine had 81 registered JI projects, and 74 letters of support had been issued.

ESCOs

Several energy service companies are operating in Ukraine, among them UkrESCO (Kiev), ESCO-Rivne (Rivne), ESCO-Zakhid (Ivano-Frankivsk), Kherson-ESCO (Kherson). The largest of them are UkrESCO and ESCO-Rivne. Both of them have been created with the support of international financial institutions and organizations (EBRD in the case of UkrESCO; UNDP/GEF in the case of ESCO-Rivne). The activities of ESCOs in Ukraine are considered to be reasonably successful although it is not clear whether they will be able to remain viable without the support from grants, preferred loans and budget.

Banking sector

No direct information on the position of the banking sector regarding EE and RES projects in Ukraine has been obtained as the developing financial crisis at the time of the mission prevented interviews at the banks and at the Association of Ukrainian Banks.

An equity investment in a particular project, where an SPV is established and the state is one of the shareholders, may be possible. However, this approach is not cheap and may not be very practical for smaller projects. It is not clear whether investments in municipal projects (e.g., renovation of a CHP) may be viable in terms of cash generation for the project, mainly because of the problem of setting energy tariffs.

Establishing a new ESCO could not be viewed as a feasible option for the Investment Fund based on the experience of the existing ones. However, investments in the already established and active ESCOs could be possible.

One of the possibilities for investments is to involve co-financing partners, including IFIs, such as EBRD and the World Bank. However, this option should be explored at the level of the headquarters of those organizations rather than country offices.

A relatively large potential for renewables in Ukraine has been identified. However, this should be treated with caution: in the current situation, spending budgetary money to subsidize tariffs for renewables may not be affordable for the Government. Another problem is the lack of adequate infrastructure (transmission lines, substations) to transport the generated energy to customers, which would also require large investments.

Energy efficiency and renewable energy sources project development and finance capacities

Existing and prospective EE and RES projects

In addition to the projects and programmes described above, a number of other projects are under way.

The World Bank participates in financing some JI projects by buying about 1 million emission reduction units for the Netherlands European Carbon Facility. The World Bank supports setting up a so-called Green Investment Scheme in Ukraine in addition to a JI mechanism and is ready to purchase Ukrainian assigned amount units. The World Bank (possibly with other international carbon funds) intends to promote GHG emission reduction projects in Ukraine after 2012.

EBRD is providing a $20 million sovereign loan to be disbursed to UkrESCO to finance energy efficiency projects designed and implemented by the company. EBRD plans to provide funds for reducing GHG emissions from large government and private industrial and energy enterprises but also from municipal gas and heat suppliers. It intends to promote climate protection projects in Ukraine more through loans and possibly direct investments as well.

In particular, the Agreement envisages financing of:
- Power industry: implementation of a range of projects of high-voltage transmission lines, including Rivne NPP – Kiev line of 750 kV, the line of 330 kV in Odessa Oblast and Western Ukraine, construction of Kanev pumped storage hydro station of 500 MW capacity and the project of distribution network reconstruction in the Crimea.
- Oil and gas complex: implementation of projects with the Ukrainian companies engaged in oil and gas production and transportation (NJSC Naftogas of Ukraine, SE Ukrtransgas, and JSC Ukrtransnafta), from the installation of new efficient compressor stations and reconstruction of underground gas stores to construction of reservoirs in oil terminals.

- Municipal infrastructure: implementation of projects in district heating, wastewater and improvement of energy efficiency based on municipal guarantees in Kiev and other several large and medium-sized Ukrainian cities.

A number of municipal heat supply companies, members of the Intersectoral Association of Development of Heat Supply Systems Ukrteplokomunenergo have developed energy efficiency projects of various scales. For example, the municipal enterprise Kharkiv Heat Supply Network (Kharkiv) has developed a major infrastructure project for potential financing from the World Bank on full process. It includes production and supply of combined heat and power and decrease in the length of the network. The total required investment for the project is $200 million. A project in Zaporizhzhya has been implemented with Zaporozhstal (major steel manufacturer) on supply of excess heat (hot water) to two city neighbourhoods in summertime.

Assessment of investment project development skills

Public sector

Public sector officials would benefit from information on financial engineering and business planning basics that would provide clear and concise information on what are the general requirements of the international financial institutions for an investment project. This would provide understanding and support at the decision-making level for the identification, selection and development of energy efficiency and renewable energy investment projects and preparation of bankable project proposals.

Private sector

Interviews with companies' representatives show that there are significant disparities in financial skills and knowledge between specialists within the country. Some companies have successfully prepared proposals for major energy investment projects, particularly with IFI. Others may have viable projects but their skills to prepare bankable projects are insufficient.

Assessment of equity and mezzanine financing business development skills

The provision of mezzanine debt is closely related to the availability of senior debt. The development of skills related to the use of mezzanine debt is closely related to the availability of knowledge and skills to prepare and present a viable business plan with an addition – use of mezzanine debt to close any gaps in the project's financial plan. There is little knowledge of the use of subordinated or mezzanine debt outside the specialized institutions, such as banks and financial institutions. Training on the use of such instruments can be done simultaneously with the general training on business planning.

The following main training programmes should be considered:
- General business plan preparation training
- Training for decision-makers
- Training on mezzanine debt and specialized instruments

Investor interest

Public sector

There is a declared interest on the part of the Government in increased energy efficiency and development of renewable energy sources, which became even more pronounced after the recent gas dispute with the Russian Federation. However, it is not clear, particularly in the current financial climate, whether significant investments could come from the State Budget. Government officials in several ministries and government agencies expressed overall support for the Investment Fund but no clear interest was expressed in possible participation of the Government as an investor in the Fund. However, co-financing of some projects could in principle be discussed.

Private sector

No interest in becoming an investor in the Fund has been expressed by domestic companies.

Potential partners for project co-financing

IFIs, in particular EBRD, could be interested both in project co-financing and in becoming an investor in the Fund. The EBRD country office may be the appropriate entity for discussions on co-financing, however decisions in principle on the EBRD role as an investor, as well as on co-financing of major projects would have to be taken at the level of Headquarters.

BUSINESS DEVELOPMENT COURSE PROGRAMME

The aim of assessment missions was to determine the local capacity and training needs of local experts to prepare investment projects, to appraise their knowledge of equity participation in projects, third party finance capacities, energy service companies (ESCOs), and performance contracting. In particular, the assessment was intended to determine whether experts identified by the National Participating Institution in each country have the following skills:

- Financial engineering and business planning skills to identify, select and develop energy efficiency and renewable energy investment projects and prepare bankable project proposals;
- Business development skills to prepare the equity and/or mezzanine finance participation of an investment fund in local companies, manufacturers of energy efficient technology, energy service company; to structure and launch a third party finance company using performance contracting; and to prepare the equity participation in renewable energy projects;
- Full range of skills described above and the capacity to develop an indicative project pipeline to a standard project identification format.

This chapter sets out the key elements of an energy efficiency business development course programme based on an appraisal of the level of local skills in participating countries. It describes the recommended content, structure, timetable of classroom instruction, type of homework assignments, and the content of course materials for the capacity building courses.

Assessment of Local Skills

The international experts participating in assessment missions agreed that there is no need for a separate training on equity and mezzanine finance, as it is only one possible financial scheme of the overall financial package, along with senior debt and other possibilities that exist for a good bankable project proposal. Furthermore, the in-depth work of structuring the mezzanine is more likely to be done in collaboration with the fund managers than by the project proponent alone. Therefore the training programme should present comparative possibilities, constrains and advantages of different financial schemes, including equity and mezzanine finance.

Prior to starting the interactive Financial Engineering/Business Planning (FE/BP) Programme, it is recommended to arrange an Information Seminar for top management and decision makers in companies, facilities and utilities with potential investment projects for energy efficiency (EE) and renewable energy (RE) (projects that could be developed during the FE/BP Programme), government representatives, project developers and National Coordinators (NCs) and National Participating Institutions (NPIs). The main purpose of this seminar is awareness raising and promotion of the subsequent FE/BP training programme.

Furthermore, arranging additional seminars for policymakers in some of the participating countries would facilitate faster market formation and more efficient implementation of EE and RE projects.

Financing Engineering / Business Planning Course Programme

This section describes the recommended content, structure, timetable of classroom instruction, type of homework assignments, and the content of course materials for the training on "Financial Engineering/Business Planning".

The capacity building programme should be interactive combining classroom lectures, instructions, discussions and consultations with development and preparation of bankable project proposals (first version of Business Plan) as homework between and after the classroom sessions ("learning by doing").

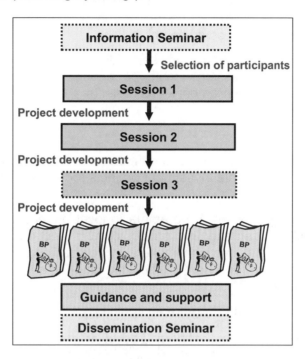

It is also possible to organize the capacity building programme without the Information Seminar. This will require additional efforts of the NCs and NPIs to market the programme, and identify and select the right participants.

Based on predefined and announced criteria, participants are selected and invited to the interactive capacity building programme, bringing suitable projects for which they will develop bankable project proposals.

Projects to be developed could either be of the type suitable for a pipeline of projects financed by an ESCO or a larger EE or RE project. For the latter type of projects, the in-depth work of structuring mezzanine financing is more likely to be done in collaboration with the fund manager (after the capacity building programme).

To further increase awareness and promote the developed projects, a Dissemination Seminar would be instrumental. In addition to the participants of the interactive capacity building programme, the target group would include top management and decision makers in companies, facilities and utilities with potential investment projects for energy efficiency and renewable energy, government representatives, financial institutions, project developers, NCs and NPIs. The main purpose of the Dissemination Seminar is raising awareness through showing outcomes of the training sessions: presentation of Business Plans developed in the course of the Programme demonstrating various ways of project financing and presentation of the new EE Investment Fund and other relevant financing facilities. The Dissemination Seminar would also give project developers an opportunity to receive feed-back on their business plans from financial institutions that will be invited.

The Dissemination Seminar could be arranged together with the third training session, or as a separate event several weeks after the last training session to ensure that the business plans are completed. By organizing the Dissemination Seminar later, the EE Investment Fund and/or other financial institutions will have time to analyze the developed

Business Plans and provide valuable comments and recommendations at the seminar. It might be useful to organize a regional Dissemination Seminar for participants from several countries if the schedule of the training sessions permits.

Additional guidance and support may be needed to the most promising projects to secure project financing. This could include further consultation on how equity or mezzanine financing from the investment fund could be utilized.

With a thorough preparation and selection process and good remote expert consultations, two classroom sessions could be enough. However, three sessions are recommended to ensure well prepared bankable projects.

Programme Outline

The purpose of the Programme is to contribute to increased awareness, knowledge and skills in preparation of bankable investment projects for energy efficiency and renewable energy.

The first target group for the Programme is local project developers, experts (e.g. ESCO staff) and managers representing project owners and utilities. The second target group is top management and decision makers in companies, facilities and utilities with potential investment projects for energy efficiency and renewable energy, government representatives and NCs and NPIs.

The expected output of one Programme could be:

Description	Indicators
Local top management/decision makers, government representatives and NCs and NPIs informed about the requirements for bankable project proposals to be presented to financial institutions	Fifty local top management/decision makers, Government representatives, and NCs and NPIs participated in the Information Seminar (and fifty at the Dissemination Seminar, if arranged)
Local consultants/experts and managers trained to identify and develop investment projects	Twenty local consultants/experts and managers completed the capacity building programme
Draft pre-feasibility studies (first version of Business Plans) ready for presentation to Financial Institutions	Eight pre-feasibility studies (first version of Business Plans) completed
Further guidance and support provided to suitable projects	Financial Institutions confirmed interest in discussing financing with three projects

The programme will include the following main activities:
- Preparations, including invitations to Information Seminar;
- Arrangement of Information Seminar;
- Selection of participants for the interactive capacity building programme (three sessions);
- Preparation of Session One;
- Implementation of Session One;
- Follow-up to Session One, evaluation of draft BP and preparation of Session Two;
- Implementation of Session Two;
- Follow-up to Session Two, evaluation of draft BP and preparation of Session Three;
- Implementation of Session Three;
- Follow-up to Session Three, evaluation of draft BP;
- Preparation of Dissemination Seminar;

- Arrangement of Dissemination Seminar;
- Providing guidance and support to further develop selected projects.

During the three classroom sessions, two days each, international trainers will give lectures, organise exercises and hold individual consultations with each team developing Pre-feasibility Study/Business Plans for specific projects. A Pre-feasibility Study/Business Plan template with the following contents is presented, and the participants trained in how to develop and present each of the chapters:
- Executive Summary;
- Borrower;
- Project Information;
- Environmental Benefits;
- Market;
- Financing Plan;
- Financial Projections;
- Project Implementation.

The participants are provided with the following training materials:
- Pre-feasibility Study/Business Plan Template;
- Business Planning Manual (introduction/description of all chapters of the Pre-feasibility Study/Business Plan);
- Exercises;
- Software tools for economic and financial calculations (profitability, disbursement plan, repayment plan, cash flow, etc.) and for environmental calculations;
- Criteria and requirements of the Investment Fund and other relevant financial institutions.

Between and after the sessions the participants continue the development of their Pre-feasibility Study/Business Plan based on the skills obtained and advice given at the previous session:

Chapter	Homework 1	Homework 2	Homework 3
Executive Summary			√
Borrower	√	√	√
Project Information	√	√	√
Environmental Benefits	√	√	√
Market	√	√	√
Financing Plan		√	√
Financial Projections		√	√
Project Implementation		√	√

The Programme can be implemented according to the following schedule; milestones are indicated as months after its beginning (international trainers contracted beforehand):

No.	Milestones	Month
1	Planning completed, programme requirements clarified with NCs and NPIs	+1
2	Information Seminar completed	+3
3	Participants selected	+4
4	Classroom Session One completed	+4
5	Classroom Session Two completed	+6
6	Classroom Session Three completed	+7
7	Pre-feasibility Study/Business Plan completed	+8
8	Dissemination Seminar completed	+9
9	Guidance and support provided to selected projects	+12
10	Programme completion report submitted	+12

There should be 4-6 weeks between the sessions to ensure that the participants have sufficient time to develop their projects properly. The Dissemination Seminar could be organized together with the third training session.

Selection of Participants

NCs and NPIs will assist in identification and invitation of participants to the Information Seminar and the Dissemination Seminar.

The content of capacity building programme will be announced at the Information Seminar. The criteria and requirements for suitable projects and for participation in the Programme include the following:
- Project design has been prepared and technical and economic evaluations have been done (e.g. Energy Audit Report);
- Sufficient equity shall be available for the project;
- Ability for each participant to dedicate around 300 working hours to the Programme, most of it related to homework between the sessions;
- Commitment to complete the homework (Business Plan) in agreed time, preferably in English (in exceptional cases a Russian version could be accepted) - Agreement to be signed before the beginning of the Programme.

All interested participants will need to complete and submit a Project Identification Form to the NCs and NPIs. In addition, a participant profile (CV) will be requested.

Information Seminar

One-day Information Seminar aims at providing clear and concise information on the general requirements of the international financial institutions to an investment project. Adequate information on possible financial instruments could assist in utilisation of new instruments, which may be employed by the fund, such as mezzanine financing, taking equity in special purpose vehicles (SPVs), or other non-recourse finance mechanisms. Furthermore, the goal of the Information Seminar is to promote the subsequent capacity building programme and invite those interested to submit an application.

This would provide understanding and support on managerial and decision-making level to the identification, selection and development of EE and RE investment projects and preparation of bankable project proposals. The agenda will also include a presentation of the new EE Investment Fund design and procedures, including:

- The possibilities offered by the fund;
- Difference from other existing schemes;
- The impacts (on the banks' evaluation procedures, senior debt access, etc.) and the implications (on corporate governance, etc.) of a mezzanine scheme;
- Financial and legal obligations.

The agenda for the Information Seminar will be tailored to each country.

Capacity Building Programme

The main subjects to be presented at the classroom sessions are listed below.

<u>Session 1:</u>

A. Programme introduction.

B. Presentation of participants.

C. Financial engineering and business planning: Introduction to the process of initiating and developing a business plan.

> Decision makers in banks have only a limited time to examine a proposed project before making their decision whether to proceed with the loan approval process. A badly presented or incomprehensible proposal stands the risk of being turned down without being fully examined. In order to overcome this problem, this seminar is designed to help potential project sponsors[11] understand how to write a business plan and how to present it.

D. Profitability calculations (Payback, Pay-off, Net Present Value, Internal Rate of Return, calculations in nominal and real terms, and exercise and introduction of software tool).

E. Project information: how to describe your project in a business plan.

> If a project is at a very early stage and it is not yet completely clear how the project is to be structured, it can be both slow and potentially error-prone to begin by developing detailed plan for the project. Clarifying the overall strategy for the project before developing the detailed plan can be a very useful technique.

F. Economic, social and environmental benefits, including exercise and introduction of software tool.

> The session provides a description of potential benefits from proposed project to the local and national economy and to the environment. Some of the benefits that come from the project can be directly quantified in money terms, such as savings in raw materials, fuel savings, reduced labour, etc. and some may be only indirectly quantifiable, including improved product quality or marketability, and which might produce a benefit in terms of increased sales. There may also be other benefits which are not quantifiable in money terms, but may have a bearing on the project. These include aspects such as safety, improved working conditions and environmental benefits.

---

[11] Project sponsors mean project owners and/or project developers. Hereinafter, the terms will be used interchangeably.

G. Measurements and verification: introduction of the International Performance Measurement and Verification Protocol (IPMVP, Volume I for energy efficiency projects, Volume III for renewable energy projects).

The implementation of the IPMVP will help to increase energy savings, document financial transactions and enhance financing for energy efficiency projects. Using this protocol will increase the confidence level of investors and project sponsor knowing that projects performance will be measured, monitored and verified according to an internationally recognized protocol.

H. Financing plan, including exercise and introduction of software tool to calculate disbursement and repayment plan.

The financing plan sets out how the transaction costs will be met. Normally, the bank will be only one of several sources of financing. In fact, the bank will require the project sponsor both to provide the equity for the project and to identify other potential sources of financing.

I. Financial schemes and sources, including equity and mezzanine financing, third party financing and performance contracting, ESCOs, etc.

The major types of finance include:
- Project Owner's Own Resources: The bank will require cash of at least 5-20 per cent of the project costs. The valuation of the in-kind contributions should be the actual current market value (resale value) and not historical costs;
- Supplier: the supplier may extend credit for the purchase of necessary materials
- Local loans: for example, these loans may come from local banks or consumer credit institutions;
- Foreign loans: these generally include loans from international financial institutions such as the World Bank and the EBRD as well as international commercial banks;
- Foreign equity: cash from other investors;
- Others: these may be grants, cash contributions or new financial instruments like mezzanine debt, performance contracting, third party finance, or combination of debt and equity, such as convertible bonds.

J. Criteria and requirements of financial institutions.

These include the basic requirements of the financial institutions to project documentation, technical, economic, financial and environmental viability of the project, its size, level of project sponsor support and enforceability of contracts.

K. Homework introduction.

Session 2:

A. Plenary presentation of homework.

B. Market, how to describe in the Business Plan.

The purpose of this section is to describe the market in which the company operates, its general characteristics, customers, competitors and factors affecting the growth of the market and the position of the company on the market.

C. Project and company cash flow, including exercise and introduction of software tool.

D. Financial projections, including cash flow, sensitivity and financial analysis.

> The project sponsor's principal aim is to prove the financial viability of the project. The bank's major concern is for the cash flow from the project to be sufficient to cover the total debt service (all payments of interest and balance of the loan). The strength of the cash flow indicates the financial viability of the project.

E. Risk assessment; evaluation of the main relevant risk elements for each project.

> In writing a business plan, the project sponsor must address specific risks of the project with the aim of presenting a clear plan of how to overcome these risks, either by mitigation or by laying them off to other parties. Each project will have different kinds of risks and the magnitude of risks will differ from project to project. However, there are key areas of risk which every project sponsor should be aware of and should keep in mind when planning to write a business plan. These key areas of risks include: borrower/sponsor risk, pre-completion risk, completion risk, technology risk, input or supply risk, operating risk, approvals risk, regulatory and environmental risk, off-take and sales risk.

F. Project implementation and management: how to organise project implementation and verification, roles and responsibilities, tendering procedures, time schedule.

> The bank will rely on the project sponsor to implement the project directly or to appoint contractors to implement the project in a timely manner and in a cost-effective way. In order for the bank to assess the risks related to implementation of the project, the sponsor will be required to describe the arrangements for implementation which include description of the major components of the project, the names of the contractor in charge of each component and the reasons for selecting a particular contractor and its relevant track record.

G. Individual consultations on each project/business plan.

H. Homework introduction.

<u>Session 3</u>:

A. Plenary presentation of homework.

B. Presentation techniques; how to present your project in a clear and understandable way to the financial institutions/investors.

C. Financial schemes and sources tailored to meet the needs of the participants and the projects being developed.

D. Individual consultations on each project/Business Plan.

E. Homework introduction.

Additional Seminar for Policymakers

The Financing Energy Efficiency Investments for Climate Change Mitigation project will also provide assistance and support to national and regional authorities to introduce economic, institutional and regulatory reforms needed to support investments in energy efficiency and renewable energy projects in the participating countries.

To support the development of these reforms and improve the financial environment in selected countries it would be beneficial to increase the awareness and knowledge of the basic requirements for bankable projects among policymakers.

This additional seminar will aim at providing clear and concise information on what are the general requirements of the international financial institutions to an investment project and what are the main economic, institutional and regulatory barriers.

The seminar for policymakers could be arranged as an add-on to the session of the Financial Engineering/Business Planning Programme. In some countries it might be useful to arrange a second (follow-up) session to organise discussions and consultations on how to overcome the main barriers in that specific country.

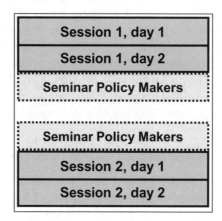

The seminar may include short presentations on the following subjects:
- Financial engineering and business planning – introduction
- Project information – how to describe your project in a Business Plan
- Financing plan
- Financial schemes and sources (including equity and mezzanine financing, third party financing and performance contracting, ESCOs, etc)
- Environmental benefits
- Market, how to describe in the Business Plan
- Financial projections (including cash flow, sensitivity and financial analysis)
- Risk assessment
- Project implementation and management
- Measurements and verification
- Criteria and requirements of financial institutions
- Main economic, institutional and regulatory barriers for investments in energy efficiency and renewable energy projects

CONCLUSIONS AND RECOMMENDATIONS

The analysis of the existing context for financing energy efficiency and renewable energy projects in the participating countries based on the assessment missions confirmed that particular conditions in each country need to be taken into consideration for a better implementation of the UNECE project Financing Energy Efficiency Investments for Climate Change Mitigation. However, a number of main barriers to successful financing of energy efficiency and renewable energy investments are common and need to be addressed in all participating countries, although at different levels.

The conclusions and recommendations reflect the preliminary assessment and analysis of the situation in the participating countries in the following areas:

- **Financial environment** for implementation of energy efficiency (EE) and renewable energy projects, including legal, regulatory and policy framework and **major barriers** to financing energy efficiency and renewable energy projects;
- Energy efficiency and renewable energy **investment project development capacities** of local experts;
- Public and private sector **investor interest** in the Eastern European Energy Efficiency Investment Fund;
- **Capacity building needs** in the countries of the region for successful development of bankable project proposals in the area of energy efficiency and renewable energy sources (RES).

Financial Environment in Energy Efficiency and Renewable Energy

Legal, regulatory and policy framework

All countries in the project region have developed or are in the process of developing **legislation, policies and strategies, which support energy efficiency and renewable energy** although levels of advancement of such legislation are very diverse among the participating countries. Many of the countries are working on amendments and improvements of existing legislation. In Albania, Kazakhstan and the Russian Federation, among others, laws on energy efficiency and/or renewable energy are under development. In Serbia, the existing Energy Law is being amended with special provisions on energy efficiency. In the former Yugoslav Republic of Macedonia, a new Energy Strategy on Renewable Energy is being developed. In the Republic of Moldova, the Energy Strategy until 2020 adopted in 2007 incorporated provisions of the draft National Programme for the Development of Renewable Energy Sources. Croatia is modifying its energy legislation to harmonize it with relevant EU Directives.

However, in many countries even when the appropriate laws exist there are problems with their **implementation**. In some cases the main reason is absence or weakness of secondary legislation (by-laws, regulations, norms and standards). Lack of enforcement of the existing legislation is another major problem. Energy strategies in a number of countries set high goals for achieving energy efficiency and increasing share of renewable energy but lack practical tools, economic incentives and appropriate funding mechanisms for achieving them. Lack of political will to implement the existing laws and strategies has also been noted.

All countries have **ratified the Kyoto Protocol** to the United Nations Framework Convention on Climate Change. In some countries, Joint Implementation (JI) or Clean Development Mechanism (CDM) projects are currently being submitted to the Designated National Authority (DNA), in other countries the DNA has been established or is being established. While the JI and CDM projects are of interest for UNECE as they are intended to mitigate climate change, they are in general not sufficiently profitable to be considered by the future Investment Fund.

For the countries of South-Eastern Europe, ratification of the Treaty establishing the Energy Community constitutes an important driving force for improving the energy efficiency environment. The **Energy Community Treaty for South-East Europe** has been signed between the European Union (EU) and the Republic of Albania, the Republic of Bulgaria, Bosnia and Herzegovina, the Republic of Croatia, the former Yugoslav Republic of Macedonia, Romania and the Republic of Serbia among the participating countries. These countries are also members of the **Task Force on Energy Efficiency** of the Energy Community Secretariat. The Task Force entails identification of the EU legislation on energy efficiency that could be extended to the member countries of the group, proposes immediate actions that would allow improvements in the energy efficiency of the region and suggests concrete measures to monitor the evolution.

Barriers to financing EE and RES projects

Energy, industry and residential sectors in all countries of the region have significant energy efficiency potential. Attracting resources to finance potential energy efficiency and renewable energy projects is regarded as an important issue by policymakers and all countries have expressed a strong interest in receiving financing (including equity and mezzanine financing) for implementation of EE and RES projects. However, several major barriers preventing financing such projects in a "business as usual" model have been identified during the assessment missions:

- **Lack of political will** by governments to provide adequate legislative and regulatory environment conducive to implementation of EE and RES projects. In many countries, while energy efficiency and development of RES is listed among the country priorities in strategic and policy documents, in practice the measures, incentives, regulation and institutional setting are insufficient for the objectives to be achieved.

- **Lack of awareness about economic benefits** from energy efficiency and renewable energy projects at all levels: government (national, regional and local), private sector, general public. Energy efficiency improvement is not perceived as a significant economic resource or potential source of revenues. Lack of supportive legislation and regulations related to public sector, which receives budget funding, results in absence of interest from regional and local authorities: due to the budgetary regulations, they often cannot keep the financial benefits of energy savings achieved. Similarly, lack of strict regulations for inefficient energy use and lack of economic incentives for improving energy efficiency make companies little interested in implementing energy efficiency projects. The following measures can be recommended to raise awareness of energy efficiency and increase interest in implementation of EE and RES projects: organization of regular information campaigns, introduction and enforcement of regulations related to energy efficiency of new buildings, power stations equipment, etc.; reform of the system of regional and municipal budgets to allow local authorities to keep savings gained from energy efficiency improvements within the community; facilitation of public tendering procedures; stimulation of metering in district heating systems; putting in place other incentives for foreign and domestic investments in energy efficiency and renewable energy.

- **Energy tariffs** (electricity and heating) are often below the market price, in particular for the residential sector. Numerous examples have shown that even under current structure of energy tariffs many EE projects can be profitable. However current energy tariffs do not allow implementation of energy savings and energy efficiency measures to be considered as economic necessity by companies, organizations and households. As a result, companies are less interested in developing adequate projects that potentially can bring significant economic benefits. In some countries, the price difference is compensated to the energy companies, but this subsidy is readjusted when the company makes savings from energy efficiency measures, which leaves little incentive to conduct energy efficiency projects.

- In all countries – whether producers or importers of primary energy sources – renewable energy is regarded as a valuable energy source. In Albania renewable energy (hydropower) covers almost 100 per cent of electricity production. In Bosnia and Herzegovina, Romania, and Serbia, hydropower provides about one third of generated electricity. However, **stimulating conditions for renewable energy development** have not been fully established as yet. Administrative obstacles for receiving permits and concessions, absence of feed-in tariffs for electricity generated from renewable energy, uncertain access to the grid do not facilitate renewable energy development.

- For a financial institution, transaction costs related to a project evaluation are of a similar level of magnitude irrespective of the size of the requested investments and of the expected financial returns. Therefore, financial institutions would usually target projects of at least several million dollars of investments. However, **most of the projects discussed with mission teams were of a smaller size then the average size considered by investment funds and banks**. Most large-scale projects introduced to mission teams could be found only in big energy companies or sometimes in the field of renewable energy, which leaves outside a number of areas strongly in need of energy efficiency measures, such as public buildings renovation, modernization of electricity production units and distribution lines or modernization of district heating.

- In this respect, **bundling several projects of a medium or small size** is crucial for enabling the future Investment Fund to operate within these sectors, often considered as a priority by the Governments of the participating countries. However, very few bundling possibilities are already in place in the countries. There are a few energy service companies (ESCO) operating in the region of the UNECE project, such as HEP-ESCO (Croatia), BelinvestESCO (Belarus), UkrESCO and RivneESCO (Ukraine). So far they have very limited experience with bundling small or medium size projects but in principle they could be used as a vehicle for such activity. Possibility of bundling such projects should be explored in the process of implementation of the UNECE project, otherwise many areas seen as national priorities by the participating countries would be excluded from attracting financing.

- In some countries (e.g. Albania, Republic of Moldova), **lack of available equity** for potential projects is a barrier to attracting additional investments in form of mezzanine and senior debt. These forms of debt are generally provided only after a reasonable percentage of equity has been raised to cover the project's expenses. The future Investment Fund may consider providing equity financing, however the issue of project ownership should be carefully considered in this case.

- **Lack of capacity for preparing bankable project proposals** is one of the major obstacles to attracting investments and needs to be addressed as early as possible through the relevant training, including Business Development Course Programme developed in the framework of this report as well as through other available tools and programmes.

The above mentioned barriers emphasize the importance of strategic government support for dissemination of information on economic benefits related to energy efficiency and renewable energy sources use and for adoption of the most needed legislative and regulatory measures that enable investments in such projects to be profitable.

Assessment of skills for preparing bankable EE and RES project proposals is one of the major objectives of this report. Assessment was made on the level of skills both in the public sector (ministries and other Government bodies at the national level, as well as regional and local authorities and state-owned organizations) and the private sector (companies in energy, industry and other sectors and banks).

The engineering and technical skills in all participating countries are sufficiently high as evident from meetings and interviews with Government officials, representatives of companies, academic institutions and other organizations. Engineering capacity building is not in the scope of this project and in general there is no need for such programme.

However, all other levels of developing a bankable EE or RE project proposal would strongly benefit from a capacity building programme addressed to project developers. An adjusted version of a capacity building programme geared mostly to awareness raising and identification of potential projects would be beneficial to policy and decision makers at the national, regional and local levels, local authorities.

Several components are necessary to develop a successful energy efficiency or renewable energy investment project:

- **Energy audits** are an essential tool to identify main sources of potential energy savings, to assess the level of possible energy savings and to recommend a first set of energy saving measures. They are considered an important first step for the preparation of an energy efficiency project. However, regulations on energy audits are missing in a number of countries and, where they exist, very often they are not mandatory or are not implemented for various reasons (e.g. lack of incentives and/or enforcement, unclear responsibility for the implementation of the regulations). There are companies and individuals who are certified to conduct energy audits work in participating countries and conduct energy audits at a professional level. However it is recognized that the number of energy auditors is insufficient. Many companies and organizations that might benefit from an energy audit are either not aware of its benefits or consider it too expensive. Energy audits could provide more precise information on particular sectors where energy savings can be achieved most efficiently. Raising awareness of the energy audits benefits, stronger regulations and mandatory character of their implementation are strongly recommended to the policymakers in the countries of the project region.

- **Measurement and verification** procedures allow to assess the exact level of energy savings that can be brought by an energy efficiency project in comparison with a baseline scenario and to state how actual savings will be verified during the implementation of the project. These procedures are a necessary step to the assessment of financial returns of the project, based on the amount of energy saved, and allow potential investors to clearly understand how exactly savings will be made. Measurement and verification procedures can be conducted by experienced energy auditors, however they are not necessarily a part of a general energy audit, and thus need to be requested specifically within a regular audit or as a separate exercise. Knowledge of this tool in the participating countries is limited. Considering the importance of measurement and verification of energy savings for the presentation of a viable business plan, it is recommended that policymakers and business developers are informed of the International Performance Measurement and Verification Protocol (IPMVP), which is the most widely used methodology of measurement and verification with a free access to detailed guidelines. Information on the application of IPMVP have been included in the proposed Business Development Course Programme.

- A **business plan** or a **bankable project proposal** is a complete document on the basis of which potential investors make a decision on financing the project. In most countries experience in developing major energy efficiency projects according to the standards of international financial institutions, banks, other potential sources of financing is very limited or non-existent. Most major EE and RES projects in the participating countries have been financed fully or partially with the use of development financing (grants, soft loans) rather than on commercial terms. Some countries (e.g. Russian Federation, Ukraine) have more experience with financing major EE and RES projects, particularly in the capitals, some other major cities and industrial regions. However the conclusion of the report is that project developers in all countries would benefit from a business development training programme tailored to the requirements and criteria of the Investment Fund to be established in the framework of the UNECE project and in accordance with international financial standards. Information on **equity and mezzanine financing** has been included in the Business Development Course Programme as in many countries potential project developers do not have clear understanding on these types of financing.

Investor Interest

Public sector

During the assessment missions, meetings and interviews were held with Government officials, including in the Ministries of Energy, the Ministries of Environment, the Ministries of Economy and the Ministries of Finance in most participating countries. References were made to the Declaration of the Ministerial Conference Environment for Europe, held in Belgrade in October 2007, where the Ministers of Environment declared that they "welcome the project on Financing Energy Efficiency Investments for Climate Change Mitigation and … will consider participating as public-sector investors in the energy efficiency investment fund, which is being created through the Energy Efficiency 21 Project." Government representatives, particularly at the Ministries of Environment confirmed that a general commitment made in Belgrade remains valid. However, no specific commitments were made by the Governments regarding possible participation in the Investment Fund as a public sector investor. This can be attributed at least partially to limited availability of resources from the national budgets, particularly in the current difficult economic and financial situation, and to the fact that the Fund's investments will not be made exclusively within a particular country. In many cases it was stated that more information on the Investment Fund is needed before any decision about participation could be taken. For instance, such interest has been shown by Government officials in the Russian Federation and Serbia.

Private sector

When assessment missions took place, the tender process for selecting the Investment Fund designer was still ongoing and detailed information on the future Fund's operations was not ready. Several requested meetings with banks have not taken place for various reasons. In some countries, a financial crisis was quoted as a reason for lack of interest in the banking community to participate in the assessment mission meetings. Some of the approached banks considered such meetings premature while in other cases banks expressed a clear lack of interest in investing into the Fund.

The meetings were conducted with the banks and financial companies in several countries, among them: Albanian Association of Banks, Albinvest, National Commercial Bank (BKT) (Albania), Belinvestbank, Belarusbank and UniCredit Bank (Belarus), Croatian Bank for Reconstruction and Development (HBOR) (Croatia), TuranAlem (Kazakhstan), Russian Bank of Development (RosBD, Russian Federation), Macedonian Bank for Development Promotion (the former Yugoslav Republic of Macedonia). Among these banks, TuranAlem (Kazakhstan) expressed interest in considering participation as investor in the Fund and requested additional information to assess this possibility. The Investment Fund designer should follow up on this expression of interest.

Several international banks and financial companies have been visited during the assessment missions, including KfW (in Albania, Bosnia and Herzegovina, and Serbia), Raiffeisen Bank (in Bosnia and Herzegovina), Small Enterprise Assistance Funds (SEAF) (in the former Yugoslav Republic of Macedonia), Société Générale (in Albania and in Serbia). The offices in the countries were not able to make commitments and recommended to engage in further discussions with the headquarters on potential interest for investor participation in the Fund.

Among the International Financial Institutions (IFIs) working in the participating countries, strong interest and experience in financing EE and RES projects has been shown by the World Bank and the European Bank for Reconstruction and Development (EBRD). The **World Bank** mainly provides soft loans, for which financial requirements are not as strict as for an equity and mezzanine fund. It is not likely that the World Bank may become an investor in the Fund.

Meetings in a number of **EBRD** country offices (including Belarus, the former Yugoslav Republic of Macedonia, Kazakhstan, and Ukraine) indicated that EBRD may be interested in becoming an investor in the Investment Fund. Representatives of the EBRD country offices expressed strong interest in getting more information about the Fund and its future operations. However it was clearly stated that such decision can only be taken at the level of the EBRD headquarters in London. The Investment Fund designer should follow up on this expression of interest.

Potential partners for project co-financing

The above mentioned banks could be considered as potential co-financing partners for the Investment Fund. While their interest in becoming potential investors in the Fund has been limited, they expressed more interest in co-financing of potential projects in their respective countries. The Investment Fund to be created will be able to provide equity or mezzanine finance only if other co-financing entities provide senior debt. In the business-as-usual scenario, major share (approximately 70 per cent) of total needed investment in a project, including EE and RES projects, is covered by senior debt. In this respect, availability of reliable co-financing partners is of crucial importance for the success of the Investment Fund. Such partners should be aware of the economic returns that energy efficiency and renewable energy projects can generate and be willing to provide senior debt on a medium- or long-term basis.

However, the assessment missions showed that in very few cases the banks have an established practice of financing EE and RES projects. Such situation is due to, on the side of project developers:
- Lack of capacity of project owners to develop an appealing business plan;
- Small financial size of many EE and RES projects and absence of bundling schemes.

And, on the side of banks:
- Lack of awareness of economic benefits that such projects can bring

- Lack of practice to lend on a medium- or long-term basis and a clear preference for lending on a short-term basis with high interest rates.

In order to make the overall project successful, awareness of banks in the countries on the benefits of participation in EE and RES projects needs to be raised. Establishing a result-oriented dialogue between the banking community and energy efficiency practitioners should be one of the priorities of the National Participating Institutions (NPIs).

EBRD provides credit lines dedicated to EE and RES projects to selected banks in several participating countries, which makes such banks preferential co-financing partners for the Fund. In some countries (e.g. in Croatia), additional mechanisms to encourage banks to provide financing for EE and RES projects exist but are so far underutilized.

During the meetings, interest of becoming a co-financing partner has been clearly expressed by:
- Local banks: the Albanian Association of Banks on behalf of several of their members (Albania), Belinvestbank and Belarusbank (Belarus), the Macedonian Bank for Development Promotion (the former Yugoslav Republic of Macedonia), TuranAlem (Kazakhstan), Russian Bank of Development (RosBD, Russian Federation).
- International banks and corporations: KfW (Albania, Bosnia and Herzegovina), UniCredit Bank (Belarus), Raiffeisen Bank (Bosnia and Herzegovina), Small Enterprise Assistance Funds (SEAF) (the former Yugoslav Republic of Macedonia), Société Générale (Albania, Serbia).

Contacts with these banks should be followed upon by the National Coordinators and NPIs in the countries as well as by the Investment Fund designer in order to raise their awareness during the time needed for the Fund to be established and for capacity building programmes for project developers to be conducted.

Co-financing of projects by Governments is another possibility. In many countries (e.g. Belarus, Croatia, Republic of Moldova and Ukraine), various budgetary and extra-budgetary funds exist with an objective of funding projects in the areas of environmental protection, energy efficiency and renewable energy sources. In most cases, funding from this source comes as a grant or a soft loan. Some countries (e.g. Kazakhstan, Russian Federation) have investment funds, which could in principle participate as co-financing partners in the projects for the Investment Fund.

Business Development Course Programme

The Business Development Course Programme has been developed in response to the capacity building needs identified in the chapters on project development and finance capacities for each country. The proposed Programme includes Information Seminar, Project Development Course, Dissemination Seminar and optional training for policymakers.

The Programme starts with the **Information Seminar**, which will contribute to raising awareness about economic benefits of energy efficiency and renewable energy projects and about major requirements of financial institutions to investment projects. The seminar will also inform decision makers of companies with potential investment projects and project developers about the content of the course. The seminar is expected to carry the following benefits:
- Encourage policymakers to support project developers by disseminating information and, when possible and necessary, through adoption of regulations and delivery of permits and concessions;
- Encourage private sector decision makers to foster development of energy efficiency projects within their companies;

- Prepare project developers to the work that will be required from them in order to complete a business plan by the end of the Programme.

The **Project Development Course** should consist of two or three sessions aimed at training the participants in financial engineering and business planning skills to identify, select and develop energy efficiency and renewable energy investment projects and prepare a complete business plan in accordance with requirements of financial institutions.

- The **first session** will introduce measurement and verification procedures, principles of financial engineering and business planning, present software that will help emphasize the environmental benefits of the project, illustrate major types of finance that can be requested by project owners and major requirements of financial institutions. A two months time will be given to project developers between the first and the second session to develop a draft business plan to be presented and discussed at the second session. During this time, trainers should be available to project developers for consultations by electronic means on their individual projects.

- The **second session** will start with presentations made by project developers and comments by trainers and participants. Profitability calculations, project and company cash flow and financial projections will be considered in more detail. In addition, the session will provide practical information (and when necessary software tools) of how to describe the market related to the project and to the company, to assess and mitigate project risks, and to implement and manage the project. By the third session, a more detailed draft business plan should be prepared.

- The **third session** will help finalise specific details of business plans, then focus on appropriate presentation techniques needed when introducing the project to potential investors and on most suitable financial schemes and sources for each specific project under preparation. Individual consultations will help finalise each business plan.

The objective of the **Dissemination Seminar** is to further raise awareness of decision makers from the public and private sectors through a presentation of business plans developed during the Project Development Course and to allow project developers to receive feedback from financial institutions that will be invited to the session.

An optional **training for policymakers** would be able to provide clear and concise information on the general requirements of the financial institutions to an investment project as well as on the main economic, institutional and regulatory barriers that need to be addressed in order to assist project owners in leveraging necessary investments. This training could be organised in parallel with the second or the third session of the Project Development Course.

The overall objective of the Programme is to significantly increase awareness of economic benefits from energy efficiency and renewable energy projects, to raise the quality of business plans and to generate about 20 business plans that would be considered for financing by various financial institutions and would have a good chance to receive investments.

ANNEX

ALBANIA

Annual data and forecast [12]

	2003[a]	2004[a]	2005[a]	2006[a]	2007[a]
GDP at market prices (Lk bn)	683	766	837	909[b]	1,018[b]
GDP (US$ bn)	5.6	7.5	8.4	9.3[b]	11.3[b]
Real GDP growth (%)	5.7	5.9	5.5	5.0[b]	6.0[b]
Consumer price inflation (av; %)	2.4	2.3	2.4	2.4	2.9
Population (m)	3.1	3.1	3.2	3.2	3.2[b]
Trade balance (US$ m)	-1,336.6	-1,591.6	-1,821.3	-2,122.8	-2,922.6
Exports of goods fob (US$ m)	447.1	603.4	656.2	792.9	1,076.0
Imports of goods fob (US$ m)	-1,783.4	-2,195.0	-2,477.5	-2,915.7	-3,998.6
Current-account balance (US$ m)	-406.8	-358.0	-571.5	-670.8	-1,201.5
Foreign-exchange reserves excl gold (US$ m)	1,009.4	1,357.6	1,404.1	1,768.8	2,104.2
Exchange rate (av) Lk:US$	121.87	102.78	99.87	98.10	90.43

[a] Actual. [b] Economist Intelligence Unit estimate.

	2003	2004	2005	2006
FDI(Foreign Direct Investment, net inflows US$ bn)[13]	1.78	3.41	2.62	3.25

---

[12] Economist Intelligence Unit, Country Report, 2008:
http://www.eiu.com/index.asp?layout=displayIssueArticle&issue_id=1523424537&article_id=1573424542.
[13] The World Bank, Data and Statistics in Albania:
http://web.worldbank.org/WBSITE/EXTERNAL/COUNTRIES/ECAEXT/ALBANIAEXTN/0,,menuPK:301437~page
PK:141132~piPK:141109~theSitePK:301412,00.html

2007 Energy Balances for Albania [14], *in thousand tons of oil equivalents (ktoe) on a net calorific value basis*

SUPPLY and CONSUMPTION	Coal	Crude Oil	Petroleum Products	Gas	Nuclear	Hydro	Geothermal, Solar, etc.	Combustible Renewable and Waste	Electricity	Heat	Total
Production	15	658	0	16	0	258	6.3	215	0	5.2	1173
Imports	3	0	954	0	0	0	0	0	243.2	0	1200.2
Exports	0	244	0	0	0	0	0	0	0	0	244
TPES (total primary energy supply)	18	414	954	16	0	258	6.3	215	243.2	5.2	2129.
TFC (total final consumption), including:	18	0	1157	2	0	0	0	215	313.5	6.3	1711
Industry sector	11.7	0	177.5	0	0	0	0	8.8	56.2	0	254.2
Transport sector	0	0	722.7	0	0	0	0	0	0	0	722.7
Other sectors, including:	6.3	0	219	2				206.2	259.3	2	694
Residential	0	0	53		0	0	0	190	178	2	423
Commercial and Public Services	6.3	0	45	2	0	0	0	12	37	0	106.3
Agriculture / Forestry	0	0	121	0	0	0	0	4.2	6.3	0	131.5
Non-Specified	0	0		0	0	0	0	0	38	0	78
Non-Energy Use	0	0	93.2	0	0	0	0	0	0	0	93.2

[14] Energy Balances for Albania, 2005. IEA Energy Statistics
http://www.iea.org/Textbase/stats/balancetable.asp?COUNTRY_CODE=AL&Submit=Submit

Annual data and forecast [15]

	2003[a]	2004[a]	2005[a]	2006[a]	2007[a]	2008[b]	2009[b]
GDP at market prices (BRb bn)	36,565	49,992	65,067	79,231	96,087	107,552	127,967
GDP at market exchange rate (US$ bn)	17.8	23.1	30.2	36.9	44.77	49.8	57.4
Real GDP growth (%)	7.0	11.5	9.4	9.9	8.2	7.0	4.5
Consumer price inflation (av; %)	28.4	18.1	10.3	7.0	12,1	17.3	19.0
Population (mid-year; m)	9.8	9.8	9.8	9.7	9.7[b]	9.6	9.6
Trade balance (US$ m)	-1,248	-2,184	-501	-2,399	-4,418		
Exports of goods fob (US$ m)	10,076	13,942	16,109	19,838	24,275	32,336	36,955
Imports of goods fob (US$ m)	-11,324	-16,126	-16,610	-22,237	28,693	-37,245	-43,233
Current-account balance (US$ m)	-434	-1,194	434	-1,512	-2,875	-3,817	-5,062
Reserves excl gold (US$ m)	595	749	1,137	1,163	4,445	-7.7	-8.8
Exchange rate (official; av; BRb:US$)	2,051.3	2,160.3	2,153.8	2,144.6	2,146.07	2,160.0	2,230.0

[a] Actual. [b] Economist Intelligence Unit estimates

	2003	2004	2005	2006
FDI(Foreign Direct Investment, net inflows US$ bn)[16]	1.72	1.64	3.05	3.54

----

[15] Economist Intelligence Unit, Country Profile, 2008
http://www.eiu.com/index.asp?layout=displayIssueArticle&issue_id=103455395&article_id=433455428
[16] The World Bank, Data and Statistics in Belarus:
http://web.worldbank.org/WBSITE/EXTERNAL/COUNTRIES/ECAEXT/BELARUSEXTN/0,,menuPK:328457~pagePK:141132~piPK:141109~theSitePK:328431,00.html

SUPPLY and CONSUMPTION	Coal and Peat	Crude Oil	Petroleum Products	Gas	Nuclear	Hydro	Geothermal, Solar, etc.	Combustible Renewables and Waste	Electricity	Heat	Total
Production	515	1789	0	182	0	3	0	1416	0	0	3904
Imports	113	21010	1323	17254	0	0	0		873	0	40572
Exports	-83	-1144	-14157	0	0	0	0	0	-498	0	-15881
TPES	567	21395	-12397	17246	0	3	0	1416	375	0	28605
TFC	425	1583	4707	4411	0	0	0	948	2448	5906	20427
Industry sector	57	0	547	1544	0	0	0	153	1197	2048	5546
Transport sector	7	0	1673	358	0	0	0	0	174	0	2212
Other sectors, including:	350	0	2041	1257	0	0	0	795	1077	3858	9378
Residential	272	0	1253	1190	0	0	0	619	493	2449	6278
Commercial and Public Services	2	0	4	4	0	0	0	146	291	1244	1692
Agriculture / Forestry	3	0	726	29	0	0	0	29	133	165	1085
Non-Specified	73	0	58	34	0	0	0	0	159	0	324
Non-Energy Use	10	1583	446	1252	0	0	0	0	0	0	3291

[17] Energy Balances for Belarus, 2006. IEA Energy Statistics
http://www.iea.org/Textbase/stats/balancetable.asp?COUNTRY_CODE=BY

Annual data and forecast

	2003	2004	2005	2006	2007[b]	2008[c]
Nominal GDP (US$ mill) [18]	7,755	9,316	10,040	11,511	14,331	18,254
Nominal GDP (KM mill)	13,443	14,678	15,791	17,950	20,479	23,107
Real GDP growth (%)	3	6.0	5.5	6.2	5.5	5.0
Population (m)	3.8	3.8	3.8	3.8	3.8	3.9
GDP per capita (US$ at PPP)[19]	5,198	5,568	5,941	7,132[b]	7,709	8,308
Trade balance (KM mill) [20]	-5,937	-6,410	-7,397	-6,225	-7,962	-6,845
Goods: exports fob (KM mill)	2,428	3,013	3,783	5,164	5,937	5,092
Goods: imports fob (KM mill)	-8,365	-9,423	-11,180	-11,389	-13,899	-11,937
Exchange rate KM:US$ (end-period)	1.73	1.57	1.57	1.55	1.34[a]	1.26
Exchange rate KM:€(end-period) [21]	1.95	1.95	1.95	1.95	1.95	1.95
Consumer price inflation (av; %)	0.6	0.4	3.7	7.5	1.6	7.5

	2003	2004	2005	2006	2007
State Budget Expenditure (KM mill) [22]	536,314	480,568	581,346	954,599	1,013,709
FDI (Foreign Direct Investment, net inflows US$ bn) [23]		1,044	823	429 (- Sept)	

---

[18] Source: Central Bank of BiH

[19] PEEREA report of BiH

[0] Source: Directorate for Economic Planning of BiH (EPPU-Economic Policy Planning Unit)

[2] Currency board arrangement: fixed rate 1 EUR = 1.9583 BAM

[2] Source: Ministry of Finance and Treasury of BiH

[3] Source: Central Bank of BiH

SUPPLY and CONSUMPTION	Coal and Peat	Crude Oil	Petroleum Products	Gas	Nuclear	Hydro	Geothermal Solar. etc.	Combustible Renewables and Waste	Electricity	Heat	Total
Production	3271	0	0	0	0	504	0	182	0	0	3957
Imports	459	149	1058	320	0	0	0		259	0	2245
Exports	-384	0	-5	0	0	0	0	0	-441	0	-829
TPES	3363	149	1053	320	0	504	0	182	-181	0	5389
TFC	428	0	1124	260	0	0	0	182	670	91	2755
Industry sector	157	0	0	205	0	0	0	0	207	0	568
Transport sector	0	0	816	0	0	0	0	0	0	0	816
Other sectors, including:	271	0	215	55	0	0	0	182	463	91	1278
Residential	0	0	0	47	0	0	0	182	354	0	584
Commercial and Public Services	0	0	0	8	0	0	0	0	109	0	116
Agriculture / Forestry	0	0	0	0	0	0	0	0	0	0	0
Non-Specified	271	0	215	0	0	0	0	0	0	91	578
Non-Energy Use	0	0	93	0	0	0	0	0	0	0	93

[24] Energy Balances for BiH, 2006. IEA Energy Statistics
http://www.iea.org/Textbase/stats/balancetable.asp?COUNTRY_CODE=BA

Annual data and forecast [25]

	2003[a]	2004[a]	2005[a]	2006[a]	2007[a]	2008[b]	2009[b]
GDP							
Nominal GDP (US$ m)	20,021	24,679	27,260	31,690	39,551	52,185	58,685
Nominal GDP (Lv m)	34,628	38,823	42,797	49,361	56,520	65,426	74,773
Real GDP growth (%)	5.0	6.6	6.2	6.3	6.2	6.1	5.8
Population (m)	7.8	7.7	7.7	7.6[c]	7.6[c]	7.5	7.5
Exchange rate Lv:US$ (end-period)	1.55	1.44	1.66	1.49	1.51	1.25	1.31
Exchange rate Lv:€(end-period)	1.95	1.94	1.96	1.96	2.20	1.96	1.96
Consumer prices (end-period; %)	5.6	4.0	6.5	6.5	15.3	9.7	6.0
Trade balance	-2,576	-3,688	-5,491	-7,028	-10,228	-12,687	-12,396
Goods: exports fob	7,081	9,931	11,776	15,101	18,441	24,155	27,971
Goods: imports fob	-9,657	-13,619	-17,267	-22,130	-28,669	-36,842	-40,367
Current-account balance	-1,022	-1,671	-3,305	-4,961	-8,530	-10,211	-9,498
International reserves (US$ m)							
Total international reserves	6,812	9,327	8,699	11,758	17,379	19,894	22,424

[a] Actual. [b] Economist Intelligence Unit forecasts. [c] Economist Intelligence Unit estimates.
Source: IMF, International Financial Statistics.

	2003	2004	2005	2006
FDI (Foreign Direct Investment, net inflows US$ bn) [26]	2.097	2.662	4.252	5.171

---

[25] Economist Intelligence Unit, Country Profile, 2009
http://www.eiu.com/index.asp?layout=displayIssueArticle&issue_id=2003628185&article_id=443628229
[26] The World Bank, Data and Statistics in Bulgaria:
http://www.worldbank.bg/WBSITE/EXTERNAL/COUNTRIES/ECAEXT/BULGARIAEXTN/0,,menuPK:305464~pagePK:141132~piPK:141109~theSitePK:305439,00.html

SUPPLY and CONSUMPTION	Coal and Peat	Crude Oil	Petroleum Products	Gas	Nuclear	Hydro	Geothermal. Solar. etc.	Combustible Renewables and Waste	Electricity	Heat	Total*
Production	4349	27	0	374	5094	364	34	836	0	0	11079
Imports	2506	7258	1637	2608	0	0	0		98	0	14107
Exports	-2	0	-3728	0	0	0	0	-32	-764	0	-4526
TPES	7065	7270	-2168	2900	5094	364	34	803	-666	0	20697
TFC	772	1	4423	1708	0	0	33	799	2312	899	10946
Industry sector	509	1	816	932	0	0	0	135	863	318	3572
Transport sector	0	0	2707	253	0	0	0	8	36	0	3004
Other sectors, including:	263	0	331	119	0	0	33	656	1414	582	3398
Residential	251	0	25	24	0	0	0	635	800	419	2155
Commercial and Public Services	6	0	65	63	0	0	33	18	597	162	944
Agriculture / Forestry	6	0	241	32	0	0	0	3	17	0	299
Non-Specified	0	0	0	0	0	0	0	0	0	0	0
Non-Energy Use	0	0	569	403	0	0	0	0	0	0	972

[27] Energy Balances for Bulgaria, 2006. IEA Energy Statistics
http://www.iea.org/Textbase/stats/balancetable.asp?COUNTRY_CODE=BG

Annual data and forecast [28]

	2003[a]	2004[a]	2005[a]	2006[a]	2007[b]	2008[c]	2009[c]
GDP							
Nominal GDP (US$ m)	29,672	35,960	38,887	42,925	51,794[a]	64,305	67,403
Nominal GDP (HRK m)	198,951	216,995	231,348	250,590	276,841[a]	304,175	325,893
Real GDP growth (%)	5.3	4.3	4.3	4.8	5.7[a]	4.6	4.5
Population (m)	4.5	4.5	4.5	4.5	4.5	4.5	4.5
GDP per head (US$ at PPP)	11,321	12,145	13,079	14,138	15,348	16,663	18,092
Exchange rate HRK:US$ (end-period)	6.12	5.67	6.25	5.57	5.04[a]	4.75	5.01
Consumer prices (end-period; %)	1.7	2.7	3.6	2.2	8.4[a]	3.5	2.5
Trade balance	-7,905	-8,346	-9,342	-10,487	-14,401	-19,023	-20,235
Goods: exports fob	6,311	8,215	8,960	10,644	11,981	15,949	17,230
Goods: imports fob	-14,216	-16,560	-18,301	-21,131	-26,382	-34,973	-37,466
Current-account balance	-2,162	-1,875	-2,555	-3,283	-4,479[a]	-7,464	-7,618
International reserves (US$ m)							
Total international reserves	8,191	8,758	8,800	11,488	13,675[a]	15,234	15,710

[a] Actual. [b] Economist Intelligence Unit estimates. [c] Economist Intelligence Unit forecasts.

	2003	2004	2005	2006	2007	2008[d]
State Budget Expenditure (kn bn) [29]	75.44	81.26	87.32	95.37	108.45	117.33
FDI (Foreign Direct Investment, net inflows EUR bn) [30]	1.762	0.949	1.468	2.745	3.597	

[d] Plan

----

[28] Economist Intelligence Unit, Country Profile, 2009:
http://www.eiu.com/index.asp?layout=displayIssueArticle&issue_id=1643571749&article_id=193571804
[29] Economist Intelligence Unit, Country Profile, 2009:
http://www.eiu.com/index.asp?layout=displayIssueArticle&issue_id=1643571749&article_id=193571804
[30] The World Bank, Data and Statistics in Croatia
http://www.worldbank.hr/WBSITE/EXTERNAL/COUNTRIES/ECAEXT/CROATIAEXTN/0,,menuPK:301270~page
PK:141132~piPK:141109~theSitePK:301245,00.html

SUPPLY and CONSUMPTION	Coal and Peat	Crude Oil	Petroleum Products	Gas	Nuclear	Hydro	Geothermal, Solar, etc.	Combustible Renewables and Waste	Electricity	Heat	Total
Production	0	1002	0	2216	0	516	2	412	0	0	4147
Imports	680	4004	1443	920	0	0	0		715	0	7763
Exports	0	0	-1895	-731	0	0	0	-46	-231	0	-2904
TPES	624	5095	-480	2350	0	516	2	366	483	0	8957
TFC	130	0	3534	1581	0	0	0	364	1294	229	7132
Industry sector	122	0	556	527	0	0	0	63	318	48	1633
Transport sector	0	0	2004	0	0	0	0	0	26	0	2030
Other sectors, including:	8	0	655	677	0	0	0	301	950	182	2772
Residential	7	0	304	539	0	0	0	301	561	146	1857
Commercial and Public Services	2	0	130	122	0	0	0	0	383	35	672
Agriculture / Forestry	0	0	222	15	0	0	0	0	6	0	243
Non-Specified	0	0	0	0	0	0	0	0	0	0	0
Non-Energy Use	0	0	319	378	0	0	0	0	0	0	697

---

31 Energy Balances for Croatia, 2006. IEA Energy Statistics
http://www.iea.org/Textbase/stats/balancetable.asp?COUNTRY_CODE=HR

Annual data and forecast [32]

	2003[a]	2004[a]	2005[a]	2006[a]	2007[a]	2008[b]	2009[b]
GDP							
Nominal GDP (US$ bn)	30.8	43.2	57.1	80.4	103.8	141.6	168.5
Nominal GDP (Tenge bn)	4,612	5,870	7,591	10,140	12,726	17,091	20,126
Real GDP growth (%)	9.3	9.6	9.7	10.6	8.5	5.9	6.2
Population (m)	15.0	15.1	15.2	15.4	15.6	15.7	15.9
Exchange rate Tenge: US$ (end-period)	144.22	130.00	133.77	127.00	120.30	120.44	118.54
Exchange rate Tenge:€(end-period)	181.92	175.99	157.80	167.60	175.67	186.69	174.26
Consumer prices (end-period; %)	6.8	6.7	7.5	8.4	18.8	15.5	10.8
Trade balance	3,679	6,785	10,322	14,642	15,141	26,122	23,891
Goods: exports fob	13,233	20,603	28,301	38,762	48,349	69,420	77,140
Goods: imports fob	-9,554	-13,818	-17,979	-24,120	-33,208	-43,299	-53,250
Current-account balance	-273	335	-1,056	-1,797	-7,184	825	-3,831
International reserves (US$ m)							
Total international reserves	4,962	9,277	7,070	19,127	17,629	24,281	30,563

[a] Actual. [b] Economist Intelligence Unit forecasts. [c] Economist Intelligence Unit estimate

	2003	2004	2005	2006
FDI (Foreign Direct Investment, net inflows US$ bn) [33]	2.092	4.157	1.975	6.143

----

[32] Economist Intelligence Unit, Country Profile, 2009
http://www.eiu.com/index.asp?layout=displayIssueArticle&issue_id=1453493530&article_id=1903493575
[33] The World Bank, Data and Statistics in Kazakhstan:
http://www.worldbank.org.kz/WBSITE/EXTERNAL/COUNTRIES/ECAEXT/KAZAKHSTANEXTN/0,,contentMDK:2
0212143~menuPK:361895~pagePK:1497618~piPK:217854~theSitePK:361869,00.html

2006 Energy Balances for Kazakhstan [34] *in thousand tons of oil equivalents (ktoe) on a net calorific value basis*

SUPPLY and CONSUMPTION	Coal and Peat	Crude Oil	Petroleum Products	Gas	Nuclear	Hydro	Geothermal, Solar, etc.	Combustible Renewables and Waste	Electricity	Heat	Total
Production	42271	65837	0	22125	0	668	0	73	0	0	130974
Imports	651	6471	1704	9279	0	0	0		358	0	18463
Exports	-12629	-58871	-3326	-12637	0	0	0	0	-286	0	-87749
TPES	30292	13436	-1885	18767	0	668	0	73	72	0	61423
TFC	5836	316	8689	12779	0	0	0	73	4219	8342	40254
Industry sector	5262	209	3037	890	0	0	0	0	2540	4079	16017
Transport sector	0	0	3806	0	0	0	0	0	181	0	3986
Other sectors, including:	0	0	1366	11889	0	0	0	73	1498	4263	19089
Residential	0	0	182	0	0	0	0	0	571	2027	2780
Commercial and Public Services	0	0	165	0	0	0	0	0	0	0	165
Agriculture / Forestry	0	0	855	0	0	0	0	0	531	0	1387
Non-Specified	0	0	164	11889	0	0	0	73	396	2236	14757
Non-Energy Use	575	107	480	0	0	0	0	0	0	0	1161

---

34 Energy Balances for Kazakhstan, 2006. IEA Energy Statistics
http://www.iea.org/Textbase/stats/balancetable.asp?COUNTRY_CODE=KZ

Annual data and forecast [35]

	2003[a]	2004[a]	2005[a]	2006[a]	2007[a]	2008[b]	2009[b]
GDP							
Nominal GDP (US$ m)	1,980.6	2,598.0	2,988.3	3,408.3	4,394.9	5,912.8	6,596.6
Nominal GDP (Lei m)	27,619	32,032	37,652	44,754	53,354	65,041	75,861
Real GDP growth (%)	6.6	7.4	7.5	4.8	3.0	6.5	5.5
Population (m)	3.6	3.6	3.4	3.4	3.4[c]	3.4	3.4
Exchange rate Lei:US$ (av)	13.94	12.33	12.60	13.13	12.14	11.00	11.50
Exchange rate Lei:€(av)	15.77	15.33	15.70	16.49	16.62	17.00	17.37
Consumer prices (end-period; %)	15.7	12.5	10.0	14.1	13.1	12.0	10.0
Trade balance	-623	-754	-1,192	-1,591	-2,316	-2,820	-3,030
Goods: exports fob	805	994	1,105	1,053	1,361	1,840	1,940
Goods: imports fob	-1,428	-1,748	-2,296	-2,644	-3,677	-4,660	-4,970
Current-account balance	-130	-47	-248	-392	-695	-856	-660
International reserves (US$ m)							
Total international reserves	302	470	597	775	1,334	1,450	1,750

[a] Actual. [b] Economist Intelligence Unit forecasts. [c] Economist Intelligence Unit estimates

	2003	2004	2005	2006
FDI (Foreign Direct Investment, net inflows US$ bn) [36]	0.73	0.87	1.97	2.41

---

[35] Economist Intelligence Unit, Country Profile, 2009
http://www.eiu.com/index.asp?layout=displayIssueArticle&issue_id=963543681&article_id=1423543727
[36] The World Bank, Data and Statistics in Moldova:
http://www.worldbank.org.md/WBSITE/EXTERNAL/COUNTRIES/ECAEXT/MOLDOVAEXTN/0,,contentMDK:20190402~menuPK:302276~pagePK:1497618~piPK:217854~theSitePK:302251,00.html

SUPPLY and CONSUMPTION	Coal and Peat	Crude Oil	Petroleum Products	Gas	Nuclear	Hydro	Geothermal, Solar, etc.	Combustible Renewables and Waste	Electricity	Heat	Total
Production	0	4	0	0	0	7	0	77	0	0	87
Imports	78	0	641	2268	0	0	0		322	0	3309
Exports	0	0	-2	0	0	0	0	0	-20	0	-22
TPES	86	4	654	2261	0	7	0	74	302	0	3387
TFC	82	0	636	637	0	0	0	69	442	288	2154
Industry sector	3	0	12	234	0	0	0	0	96	97	442
Transport sector	0	0	278	62	0	0	0	0	7	0	347
Other sectors, including:	79	0	334	341	0	0	0	69	339	190	1353
Residential	41	0	265	245	0	0	0	0	152	127	830
Commercial and Public Services	36	0	2	81	0	0	0	0	85	51	254
Agriculture / Forestry	0	0	57	14	0	0	0	0	7	1	79
Non-Specified	3	0	10	1	0	0	0	69	95	11	189
Non-Energy Use	0	0	12	0	0	0	0	0	0	0	12

[37] Energy Balances for the Republic of Moldova, 2006. IEA Energy Statistics
http://www.iea.org/Textbase/stats/balancetable.asp?COUNTRY_CODE=MD

Annual data and forecast [38]

	2003[a]	2004[a]	2005[a]	2006[a]	2007[a]	2008[b]	2009[b]
GDP							
Nominal GDP (US$ bn)	59.5	75.5	98.9	121.9	168.5	205.2	227.3
Nominal GDP (Lei bn)	198	246	288	342	411	492	559
Real GDP growth (%)	5.2	8.5	4.2	7.9	6.0	6.8	5.0
Population (m)	21.7	21.7	21.6	21.5	21.5	21.5	21.5
Exchange rate Lei:US$ (av)	3.32	3.26	2.91	2.81	2.44	2.40	2.46
Exchange rate Lei:€(av)	3.75	4.06	3.63	3.53	3.34	3.71	3.71
Consumer prices (av; %)	15.3	11.9	9.0	6.6	4.8	8.1	5.3
Trade balance	-2,691	-4,152	-9,618	-14,836	-24,223	-36,295	-40,550
Goods: exports fob	17,618	23,485	27,730	32,336	40,318	56,666	67,266
Goods: imports fob	-20,309	-27,637	-37,348	-47,172	-64,541	-92,961	-107,816
Current-account balance	-1,674	-3,869	-8,621	-12,785	-23,017	-31,991	-31,884
International reserves (US$ m)							
Total international reserves	9,450	16,096	21,595	30,211	39,956	53,928	63,907

[a] Actual. [b] Economist Intelligence Unit forecasts. [c] Economist Intelligence Unit estimates. [d] Consolidated government budget, including local and social security budgets.
Source: IMF, International Financial Statistics.

	2003	2004	2005	2006
FDI (Foreign Direct Investment, net inflows US$ bn) [39]	1.844	6.443	6.482	11.394

----

[38] Economist Intelligence Unit, Country Profile, 2009
http://www.eiu.com/index.asp?layout=displayIssueArticle&issue_id=303554215&article_id=873554272
[39] The World Bank, Data and Statistics in Romania:
http://www.worldbank.org.ro/WBSITE/EXTERNAL/COUNTRIES/ECAEXT/ROMANIAEXTN/0,,menuPK:287320~pagePK:141132~piPK:141109~theSitePK:275154,00.html

2006 Energy Balances for Romania [40], in thousand tons of oil equivalents (ktoe) on a net calorific value basis

SUPPLY and CONSUMPTION	Coal and Peat	Crude Oil	Petroleum Products	Gas	Nuclear	Hydro	Geothermal Solar, etc.	Combustible Renewables and Waste	Electricity	Heat	Total*
Production	6555	5504	0	9556	1468	1579	18	3315	0	0	27994
Imports	2583	8425	1323	4790	0	0	0		85	0	17204
Exports	-14	0	-5393	0	0	0	0	0	-453	0	-5860
TPES	9437	14006	-3859	14604	1468	1579	18	3265	-367	0	40149
TFC	1136	0	7967	8444	0	0	13	3089	3523	1993	26165
Industry sector	1117	0	1175	3515	0	0	1	338	2088	343	8575
Transport sector	0	0	4250	31	0	0	0	1	116	0	4397
Other sectors, including:	19	0	917	4172	0	0	13	2751	1319	1650	10841
Residential	10	0	452	2547	0	0	6	2570	860	1392	7838
Commercial and Public Services	0	0	144	1596	0	0	7	0	421	237	2405
Agriculture / Forestry	6	0	159	30	0	0	0	6	38	20	260
Non-Specified	3	0	162	0	0	0	0	174	0	0	339
Non-Energy Use	0	0	1625	726	0	0	0	0	0	0	2351

[40] Energy Balances for Romania, 2006. IEA Energy Statistics
http://www.iea.org/Textbase/stats/balancetable.asp?COUNTRY_CODE=RO

Annual data and forecast [41]

	2003[a]	2004[a]	2005[a]	2006[a]	2007[a]	2008[b]
Nominal GDP (US$ bn)	431.5	591.7	764.5	988.7	1,289.6	1,714.4
Nominal GDP (Rb bn)	13,243	17,048	21,625	26,883	32,989	41,145
Real GDP growth (%)	7.3	7.2	6.4	7.4	8.1	7.5
Population (m)	144.6	143.8	143.1	142.6	142.3	141.8
GDP per capita (US$ at PPP)	9,694	10,746	11,861	13,190	14,661	16,205
Trade balance (US$ m)	59,861	85,825	118,266	139,234	132,044	162,919
Goods: exports fob (US$ m)	135,930	183,207	243,569	303,926	355,465	469,262
Goods: imports fob (US$ m)	-76,069	-97,382	-125,303	-164,692	-223,421	-306,344
Exchange rate Rb:US$ (av)	30.69	28.81	28.28	27.19	25.58	24.00
Exchange rate Rb:€(end-period)	34.71	35.82	35.24	34.14	35.01	37.08
Consumer price inflation (av, %)	13.7	10.9	12.7	9.7	9.0	13.9

[a] Actual. [b] Economist Intelligence Unit forecasts. [c] Economist Intelligence Unit estimates

	2003	2004	2005	2006
FDI (Foreign Direct Investment, net inflows US$ bn) [42]	7.958	15.444	12.885	30.827

[41] Economist Intelligence Unit, Country Profile, 2008
http://www.eiu.com/index.asp?layout=displayIssueArticle&issue_id=873520272&article_id=1363520321
[42] The World Bank, Data and Statistics in Russia
http://web.worldbank.org/WBSITE/EXTERNAL/COUNTRIES/ECAEXT/RUSSIANFEDERATIONEXTN/0,,contentM
DK:21054807~menuPK:517666~pagePK:1497618~piPK:217854~theSitePK:305600,00.html

SUPPLY and CONSUMPTION	Coal and Peat	Crude Oil	Petroleum Products	Gas	Nuclear	Hydro	Geothermal, Solar, etc.	Combustible Renewables and Waste	Electricity	Heat	Total
Production	143251	478130	0	525724	41116	14908	398	7380	0	9068	1219975
Imports	14203	2332	44	5814	0	0	0		440	0	22833
Exports	-51308	-249685	-87319	-163842	0	0	0	0	-1800	0	-553954
TPES	106737	228628	-89387	358605	41116	14908	398	7482	-1360	9068	676196
TFC	16270	59	101056	130658	0	0	0	2427	58600	122661	431733
Industry sector	10319	5	10868	31585	0	0	0	363	30406	47219	130765
Transport sector	0	12	56018	33558	0	0	0	0	7385	0	96974
Other sectors, including:	5387	42	12001	43816	0	0	0	2064	20809	75443	159562
Residential	3079	0	6459	39554	0	0	0	1245	9677	54876	114890
Commercial and Public Services	2172	34	955	3775	0	0	0	578	9666	17437	34617
Agriculture / Forestry	134	8	4123	487	0	0	0	240	1442	3118	9552
Non-Specified	0	0	0	0	0	0	0	0	0	0	0
Non-Energy Use	564	0	22169	21699	0	0	0	0	0	0	44433

[43] Energy Balances for Russian Federation, 2006. IEA Energy Statistics
http://www.iea.org/Textbase/stats/balancetable.asp?COUNTRY_CODE=RU

Annual data and forecast [44]

	2003[a]	2004[a]	2005[a]	2006[a]	2007[b]	2008[c]	2009[c]
GDP							
Nominal GDP (US$ m)	20,397	24,387	26,039	31,815	41,881	55,811	64,740
Nominal GDP (RSD bn)	1,172	1,431	1,750	2,126	2,435	2,879	3,271
Real GDP growth (%)	2.5	8.4	6.2	5.7	7.5	6.5	6.0
Population (m)	7.5	7.5	7.4	7.4	7.4	7.4	7.4
Exchange rate RSD:US$ (end-period)	54.64	57.94	72.22	59.98	53.73[a]	50.61	51.43
Exchange rate RSD:€(end-period)	68.92	78.43	85.19	79.15	78.46[a]	79.45	76.63
Consumer prices (end-period; %)	7.8	13.8	17.6	6.6	10.1[a]	9.5	6.5
Trade balance	-4,847	-7,047	-5,563	-6,271	-8,831[a]	-11,834	-13,142
Goods: exports fob	2,477	3,897	4,647	6,442	8,858[a]	11,515	13,243
Goods: imports fob	-7,324	-10,944	-10,210	-12,713	-17,689[a]	-23,349	-26,385
Current-account balance	-1,928	-2,922	-2,088	-3,967	-6,888[a]	-9,088	-10,036
International reserves (US$ m)							
Total international reserves	3,550	4,245	5,745	11,875	14,215[a]	15,000	15,500

[a] Actual. [b] Economist Intelligence Unit estimates. [c] Economist Intelligence Unit forecasts.
Source: IMF, International Financial Statistics.

	2003	2004	2005	2006	2007	2008
State Budget Expenditure (billion RSD)	391.6	355.9	438.8	529.7	617.4	441.1
FDI (Foreign Direct Investment, net inflows €m)[45]	1,205.70	776.6	1,244.60	3,398.70	1,601.60	

[44] Economist Intelligence Unit, Country Profile, 2009:
http://www.eiu.com/index.asp?layout=displayIssueArticle&issue_id=493631834&article_id=1143631899
[45] The World Bank, Data and Statistics in Serbia:
http://www.worldbank.org.yu/WBSITE/EXTERNAL/COUNTRIES/ECAEXT/SERBIAEXTN/0,,menuPK:300929~pag
ePK:141132~piPK:141109~theSitePK:300904,00.html

SUPPLY and CONSUMPTION	Coal and Peat	Crude Oil	Petroleum Products	Gas	Nuclear	Hydro	Geothermal Solar, etc.	Combustible Renewables and Waste	Electricity	Heat	Total
Production	7812	660	0	236	0	943	0	907	0	0	10558
Imports	913	2538	1557	1755	0	0	0		737	0	7501
Exports	-46	0	-68	0	0	0	0	-100	-806	0	-1020
TPES	8710	3198	1489	1991	0	943	0	807	-70	0	17068
TFC	1116	0	3724	1398	0	0	0	801	2258	988	10285
Industry sector	556	0	1052	792	0	0	0	0	561	501	3462
Transport sector	0	0	2414	0	0	0	0	0	22	0	2437
Other sectors, including:	560	0	118	214	0	0	0	801	1675	488	3856
Residential	475	0	26	214	0	0	0	801	1209	375	3100
Commercial and Public Services	0	0	0	0	0	0	0	0	447	103	550
Agriculture / Forestry	0	0	0	0	0	0	0	0	18	9	28
Non-Specified	86	0	93	0	0	0	0	0	0	0	178
Non-Energy Use	0	0	139	392	0	0	0	0	0	0	530

[46] Energy Balances for Serbia, 2006. IEA Energy Statistics
http://www.iea.org/Textbase/stats/balancetable.asp?COUNTRY_CODE=RS

Annual data and forecast[47]

	2003[a]	2004[a]	2005[a]	2006[b]	2007[b]	2008[c]	2009[c]
GDP							
Nominal GDP (US$ m)	4,629.7	5,368.5	5,767.6	6,089.7	7,004.7	8,264.8	8,497.6
Nominal GDP (Den m)	251,486	265,257	284,226	297,176	313,322	328,988	346,095
Real GDP growth (%)	2.8	4.1	4.0	4.3[a]	5.1[a]	5.0	5.2
Population (m)	2.0[b]	2.0[b]	2.0[b]	2.0	2.1	2.1	2.1
Exchange rate Den:US$ (end-period)	49.05	45.07	51.86	46.45[a]	41.66[a]	39.68	41.84
Exchange rate Den:€(end-period)	61.87	61.01	61.17	61.30[a]	60.83[a]	61.50	61.50
Consumer prices (end-period; %)	2.5	-1.9	1.2	2.9[a]	6.1[a]	7.6	5.9
Trade balance	-848	-1,112	-1,058	-1,285[a]	-1,627[a]	-2,181	-2,456
Goods: exports fob	1,363	1,672	2,040	2,396[a]	3,349[a]	4,344	4,819
Goods: imports fob	-2,211	-2,785	-3,097	-3,682[a]	-4,976[a]	-6,525	-7,275
Current-account balance	-149	-415	-81	-24[a]	-249	-697	-811
International reserves (US$ m)							
Total international reserves	935	991	1,340	1,889[a]	2,265[a]	2,389	2,569

[a] Actual. [b] Economist Intelligence Unit estimates. [c] Economist Intelligence Unit forecasts.
Source: IMF, International Financial Statistics.

	2003	2004	2005	2006
FDI(Foreign Direct Investment, net inflows US$ bn) [48]	0.96	1.57	1.00	350

---

[47] Economist Intelligence Unit, Country Profile, 2009
http://www.eiu.com/index.asp?layout=displayIssueArticle&issue_id=213524206&article_id=853524270
[48] The World Bank, Data and Statistics in FYRM:
http://www.worldbank.org.mk/WBSITE/EXTERNAL/COUNTRIES/ECAEXT/MACEDONIAEXTN/0,,contentMDK:20179336~menuPK:304498~pagePK:1497618~piPK:217854~theSitePK:304473,00.html

2006 Energy Balances for the Former Yugoslav Republic of Macedonia [49], in thousand tons of oil equivalents (ktoe) on a net calorific value basis

SUPPLY and CONSUMPTION	Coal and Peat	Crude Oil	Petroleum Products	Gas	Nuclear	Hydro	Geothermal, Solar, etc.	Combustible Renewables and Waste	Electricity	Heat	Total
Production	1129	0	0	0	0	142	10	169	0	0	1450
Imports	134	1079	246	67	0	0	0		154	0	1681
Exports	-1	0	-357	0	0	0	0	-1	0	0	-359
TPES	1254	1089	-124	67	0	142	10	166	154	0	2759
TFC	117	0	713	34	0	0	9	163	554	118	1708
Industry sector	110	0	183	33	0	0	0	3	191	57	577
Transport sector	0	0	347	0	0	0	0	0	2	0	350
Other sectors, including:	8	0	153	1	0	0	9	160	360	62	752
Residential	3	0	41	0	0	0	0	150	262	42	498
Commercial and Public Services	4	0	93	1	0	0	1	8	96	20	224
Agriculture / Forestry	0	0	19	0	0	0	8	1	2	0	31
Non-Specified	0	0	0	0	0	0	0	0	0	0	0
Non-Energy Use	0	0	29	0	0	0	0	0	0	0	29

[49] Energy Balances for FYRM, 2006. IEA Energy Statistics
http://www.iea.org/Textbase/stats/balancetable.asp?COUNTRY_CODE=MK

Annual data and forecast[50]

	2003[a]	2004[a]	2005[a]	2006[a]	2007[a]	2008[b]	2009[b]
GDP							
Nominal GDP (US$ bn)	50.1	64.9	82.9	107.8	141.2	198.6	253.8
Nominal GDP (HRN bn)	267	345	425	544	713	983	1,262
Real GDP growth (%)	9.5	12.1	2.6	7.9	7.7	6.2	5.8
Population (m)	47.4	47.1	46.7	46.5	46.2	46.0	45.8
Exchange rate HRN:US$ (end-period)	5.33	5.31	5.05	5.05	5.05	4.85	4.99
Exchange rate HRN:€(end-period)	6.73	7.18	5.96	6.66	7.37	7.52	7.33
Consumer prices (end-period; %)	8.3	12.3	10.3	11.6	16.6	18.5	11.9
Trade balance	518	3,741	-1,135	-5,194	-10,572	-16,040	-20,254
Goods: exports fob	23,739	33,432	35,024	38,949	49,840	65,679	74,671
Goods: imports fob	-23,221	-29,691	-36,159	-44,143	-60,412	-81,719	-94,925
Current-account balance	2,891	6,909	2,531	-1,617	-5,918	-12,625	-17,776
International reserves (US$ m)							
Total international reserves	6,943	9,715	19,391	22,360	32,480	35,890	39,160

[a] Actual. [b] Economist Intelligence Unit forecasts. [c] Economist Intelligence Unit estimates.
Source: IMF, International Financial Statistics.

	2003	2004	2005	2006
State Budget Expenditure[51] (HRN bn)	44,39	63,73	89,91	
FDI (Foreign Direct Investment, net inflows US$ bn)[52]	1.424	1.715	7.808	5.604

----

[50] Economist Intelligence Unit, Country Profile, 2009:
http://www.eiu.com/index.asp?layout=displayIssueArticle&issue_id=423602027&article_id=963602081
[51] Information submitted by the National Participating Institution.
[52] The World Bank, Data and Statistics in Ukraine:
http://web.worldbank.org/WBSITE/EXTERNAL/COUNTRIES/ECAEXT/UKRAINEEXTN/0,,contentMDK:20147755
~menuPK:328559~pagePK:1497618~piPK:217854~theSitePK:328533,00.html

2006 Energy Balances for Ukraine [53] *in thousand tons of oil equivalents (ktoe) on a net calorific value basis*

SUPPLY and CONSUMPTION	Coal and Peat	Crude Oil	Petroleum Products	Gas	Nuclear	Hydro	Geothermal, Solar, etc.	Combustible Renewables and Waste	Electricity	Heat	Total
Production	35339	4539	0	17685	23513	1108	3	582	0	0	82769
Imports	6745	11165	4063	42112	0	0	0		179	0	64265
Exports	-2082	-161	-4745	-3	0	0	0	0	-1077	0	-8068
TPES	40093	15485	-695	58235	23513	1108	3	582	-898	0	137427
TFC	12274	18	13571	33929	0	0	0	304	11149	11194	82440
Industry sector	8911	5	1585	10291	0	0	0	50	5906	6023	32771
Transport sector	1	0	8213	2938	0	0	0	5	851	0	12008
Other sectors, including:	2201	0	2082	15543	0	0	0	249	4393	5171	29638
Residential	1847	0	673	14880	0	0	0	172	2380	5171	25124
Commercial and Public Services	328	0	51	532	0	0	0	50	1733	0	2695
Agriculture / Forestry	25	0	1357	130	0	0	0	27	276	0	1816
Non-Specified	0	0	0	0	0	0	0	0	0	0	0
Non-Energy Use	1161	13	1692	5157	0	0	0	0	0	0	8023

[53] Energy Balances for Ukraine, 2006. IEA Energy Statistics
http://www.iea.org/Textbase/stats/balancetable.asp?COUNTRY_CODE=UA

SOURCES

Alliance to Save Energy. Energy Community Stocktaking on Energy Efficiency Report. May 2008.

Alliance to Save Energy. Urban Heating in Bulgaria: Experience from the Transition and Future Directions. 2005.

Alliance to Save Energy. Urban Heating in Moldova: Experience from the Transition and Future Directions. 2005.

ARCE. Trends in energy efficiency and renewables. 2006.

Bulgarian Energy Efficiency Agency. Bulletin of the Bulgarian Energy Efficiency Agency. 2007.

Danske Bank. Emerging Markets Briefer, 15 December 2008.

Danske Bank. New Europe Weekly, Investment Research, 22 December 2008.

Deutsche Bank. Global Markets Research, Emerging Europe Ukraine, 29 December 2008.

EBRD. Country Strategy for Croatia, May 2007.

EBRD. Renewable Energy Resource Assessment.

EBRD-World Bank. EBRD-World Bank Business Environment and Enterprise Performance Survey (BEEPS). 2005.

Economic Commission for Europe, Committee on Environmental Policy. Environmental Performance Reviews. Republic of Serbia, Second Review. United Nations, New York and Geneva, 2007.

Economic Commission for Europe, Committee on Environmental Policy. Environmental Performance Reviews. Ukraine, Second Review. United Nations, New York and Geneva, 2007.

Electric Power Industry of Serbia. Annual Report. 2007

EnEffect. Legal, organizational and technical issues related to the investment process in Bulgarian municipalities. 2003.

EnEffect. Study on the financial barriers to project implementation. Proposals for their overcoming. 2003.

ENERDATA. The energy market in Bulgaria. 2006.

ENERDATA. The energy market in Romania. 2007.

Energy Charter Protocol on Energy Efficiency and Related Environmental Aspects PEEREA. In-Depth Review of the Energy Efficiency Policy of Bulgaria. 2008.

Energy Charter Protocol on Energy Efficiency and Related Environmental Aspects PEEREA. In-Depth Review of the Energy Efficiency Policy of the Former Yugoslav Republic of Macedonia. 2007.

Energy Charter Protocol on Energy Efficiency and Related Environmental Aspects PEEREA. In-Depth Review of the Energy Efficiency Policy of the Republic of Moldova. 2004.

Energy Charter Protocol on Energy Efficiency and Related Environmental Aspects PEEREA. Regular Review of Energy Efficiency Policies of Albania. 2007.

Energy Charter Protocol on Energy Efficiency and Related Environmental Aspects PEEREA. Regular Review of Energy Efficiency Policies of Kazakhstan. 2006.

Energy Charter Protocol on Energy Efficiency and Related Environmental Aspects PEEREA. Regular Review of Energy Efficiency Policies of Romania. 2006.

EU. Screening report Croatia, Chapter 15 – Energy. June 2006.

Fejzibegovic, Semra. Mediterranean and National Strategies for Sustainable Development. Priority Field of Action 2: Energy and Climate Change. Regional Activity Centre, Hydro-Engineering Institute, Sarajevo. March 2007.

International Finance Corporation. On the road to energy efficiency: experience and future outlook. Researching energy efficiency practices among Russian companies. 2007.

International Monetary Fund. Bulgaria's credit boom: after credit limits. 2007.

International Monetary Fund. Republic of Moldova: Financial System Stability Assessment. 2005 and update 2008.

Investment Projects SPA Belenergo. Ministry of Energy of Republic of Belarus. 2008

Government of the Federation of Bosnia and Herzegovina. (2007), Programme of the FBiH Government – policy for action and basic strategies of the Government in the period 2007–2010.

Kovacic, Bojan. Fourth Poverty Reduction Strategies Forum, World Bank Institute, Athens, 26-27 June 2007.

KPMG, Power and Utilities Center of Excellence Team. Central and Eastern European Renewable Electricity Outlook. Budapest, Hungary, 2008.

National communication on Romanian first national action plan for energy efficiency (2007-2010).

RAEF. Presentation at the UNECE European Clean Energy Fund Seminar. Geneva, 21 February 2008.

Sculac Domac, Marija. Fourth International Business Forum "Investments in the Environment for a Better Quality of Life". Sofia, Bulgaria, 2 October 2007.

The World Bank. Bosnia and Herzegovina Country Partnership Strategy for the period FY08 – FY11. November 2007.

The World Bank. Improving People's Lives in Bosnia and Herzegovina, Country Brief. June 2008.

The World Bank. Moldova Investment Climate Assessment. 2004.

The World Bank. Romania Country Brief. 2008.

The World Bank. Status of projects in execution, FY07 SOPE, Bosnia and Herzegovina, October 2007.

The World Bank, Status of projects in execution, FY07 SOPE, Serbia, October 2007.

Vlada Republike Srpske. Razvojni program Republike Srpske 2007-2010. Vizija, ciljevi i

Prioriteti. Banja Luka, June 2007.

Internet Addresses

Austrian Energy Agency. Energy Profile Bosnia and Herzegovina   http://www.energyagency.at/enercee/bih/index.htm

Austrian Energy Agency. Energy Profile Croatia   http://www.energyagency.at/enercee/hr/

Austrian Energy Agency. Energy Profile Serbia   http://www.energyagency.at/enercee/sr/index.htm

CETEOR (Center for Environmental Technological Development)   http://www.pksa.com.ba/ceteor/indexe.html

Chamber of Economy of Federation of Bosnia   http://www.kfbih.com/

and Herzegovina

Chamber of Economy of Republika Srpska	http://www.pkrs.inecco.net/
Croatian Bank for Reconstruction and Development (HBOR). Environmental Protection	http: www.hbor.hr/Default.aspx?sec=1483
Croatian Bank for Reconstruction and Development (HBOR). Preparation of Renewable Energy Resources Projects	http://www.hbor.hr/Default.aspx?sec=1329
Croatian Bank for Reconstruction and Development (HBOR). Programme of Issuing Bank Guarantees	http://www.hbor.hr/Default.aspx?sec=1602
Croatian Chamber of Economy's Energy Association	http://www2.hgk.hr/en/
Croatian Energy Market Operator (HROTE)	http://www.hrote.hr/
Croatian Energy Regulatory Agency	http://www.hera.hr/
European Bank for Reconstruction and Development (EBRD)	http://www.ebrd.com
EBRD. Bosnia and Herzegovina Country factsheet	http://www.ebrd.com/bih
EBRD. Bulgaria Country factsheet. 2008.	http://www.ebrd.com/bulgaria
EBRD. Croatia Country factsheet. 2008.	http://www.ebrd.com/croatia
EBRD. Country strategy for Bosnia and Herzegovina (economy, infrastructure, financial sector, enterprise sector) (2007)	http://www.ebrd.com/about/strategy/country/bosnherz/strategy.pdf
EBRD. Serbia Country factsheet	http://www.ebrd.com/serbia
Economist Intelligence Unit	http://www.eiu.com
EETEK Holding	http://www.eetek.hu/en/index.html
Energy Institute Hrvoje Pozar	http://www.eihp.hr/
Foreign Investment Promotion Agency of Bosnia and Herzegovina	http://www.fipa.gov.ba/
Government of Federation of Bosnia and Herzegovina	http://www.fbihvlada.gov.ba/
HEP ESCO	http://www.hepesco.hr/esco/en/aboutus/default.aspx
International Energy Agency (IEA)	http://www.iea.org/
IEA. Country Statistics	http://www.iea.org/Textbase/stats/index.asp
International Finance Corporation (IFC)	http://www.ifc.org
IFC. Russia Sustainable Energy Finance Program	http://www.ifc.org/russia/energyefficiency (Russian) http://www.ifc.org/ifcext/rsefp.nsf/Content/Home (English)
International Monetary Fund (IMF)	http://www.imf.org/
IPM Research Centre	http://research.by/eng/info/

IPM Research Centre. Economy of Belarus	http://research.by/eng/data/economy
Kazakhstan Sustainable Energy Financing Facility (KazSEFF)	http://www.kazseff.kz
Ministry of Economy, Labour and Entrepreneurship of Croatia	http://www.mingo.hr/
Ministry of Energy of Belarus	http://www.minenergo.gov.by/en
Ministry of Energy of Russian Federation	http://minenergo.com/
Ministry of Environmental Protection and Physical Planning and Construction of Croatia	http://www.mzopu.hr/
Ministry of Environment and Spatial Planning of Serbia	http://www.ekoplan.gov.rs/src/index.php
Ministry of Foreign Trade and Economic Relations (MOFTER) of Bosnia and Herzegovina	http://www.mvteo.gov.ba/home/index.php?lang=en
Ministry of Fuel and Energy of Ukraine	http://mpe.kmu.gov.ua/fuel/control/uk/index
Ministry of Mining and Energy of Serbia	http://www.mem.sr.gov.yu/
Regional Economic Development Strategies	http://www.eured-smebih.org/index.php?option=com_content&task=view&id=31&Itemid=58
Republika Srpska Government	http://www.vladars.net/
Serbian Chamber of Commerce	http://www.pks.co.yu/
Serbian Energy Agency (SEA)	http://www.aers.org.yu/IndexEng.asp
Serbian Energy Efficiency Agency (SEEA)	http://www.seea.sr.gov.yu/English/Prezentacija1.htm
State Comprehensive Program of Energy System Upgrading of Belarus	http://president.gov.by/en/press20032.html
The World Bank	http://www.worldbank.org
The World Bank. Project Portfolio for Croatia	http://www.worldbank.hr